# Fixed-Film Reactors in Wastewater Treatment

# Fixed-Film Reactors in Wastewater Treatment

Nicholas F Gray
Trinity College, University of Dublin, Ireland

World Scientific
NEW JERSEY • LONDON • SINGAPORE • BEIJING • SHANGHAI • HONG KONG • TAIPEI • CHENNAI • TOKYO

*Published by*

World Scientific Publishing Europe Ltd.
57 Shelton Street, Covent Garden, London WC2H 9HE
*Head office:* 5 Toh Tuck Link, Singapore 596224
*USA office:* 27 Warren Street, Suite 401-402, Hackensack, NJ 07601

**Library of Congress Cataloging-in-Publication Data**
Names: Gray, N. F., author.
Title: Fixed-film reactors in wastewater treatment / Nicholas F Gray, Trinity College, University of Dublin, Ireland.
Description: New Jersey : World Scientific, [2021] | Includes bibliographical references and index.
Identifiers: LCCN 2020032158 | ISBN 9781786349248 (hardcover) | ISBN 9781786349255 (ebook) | ISBN 9781786349262 (ebook other)
Subjects: LCSH: Sewage--Purification--Biological treatment. | Water--Purification--Fixed-film biological process.
Classification: LCC TD475 .G73 2021 | DDC 628.3/5--dc23
LC record available at https://lccn.loc.gov/2020032158

**British Library Cataloguing-in-Publication Data**
A catalogue record for this book is available from the British Library.



For any available supplementary material, please visit
https://www.worldscientific.com/worldscibooks/10.1142/Q0271#t=suppl

Desk Editors: Gregory Lee/Amanda Yun

Typeset by Stallion Press
Email: enquiries@stallionpress.com

Dedicated to Lucy, Cat and Becky.

'To you it's just crap, to me it's bread and butter.'
Spike Milligan
*Recollections of the latrine orderly*

# Preface

In 1989 I finally published the textbook *Biology of Wastewater Treatment*. I was and have remained primarily a wastewater engineer with a strong interest in biology. At that time there was no single reference work that encompassed my areas of interest and research. The book, written during the pre-Internet era, was very much based on research literature and my own experience. It ended up being 828 pages long. When Imperial College Press published the last edition in 2004, it had grown to a rather unwieldy 1,440 pages. When approached by the publishers for a new edition, it proved just too difficult a task to achieve within the confines of a single textbook. So it was decided to break down the original book into a series of key chapters and to produce a series of smaller, more focused textbooks. This new text is the revised edition of *Biology of Wastewater Treatment* that deals with fixed-film reactors, also referred to as attached biofilm reactors. I started my career working at a Yorkshire Water percolating (trickling) filter wastewater treatment plant and fell in love with the sedate and almost silent process. The biology of the process fascinated me with its complex microbial ecology and macroinvertebrate grazers, and it actually became part of the larger ecosystem of the area with a wide variety of terrestrial visitors, including birds and mammals. Over the decades, I have seen percolating filters being slowly replaced with noisy energy-hungry alternatives, and for a while, biofilm systems seemed consigned to the past. But I am glad to say that they remain a key treatment option in industry where we see rapid development in media and design. Biological air treatment systems have become commonplace, with biofiltration and biotrickling filters increasingly replacing physical-chemical scrubbers. From the demand for more efficient, operator-friendly and sustainable treatment solutions we have seen a resurgence in fixed-film systems in municipal wastewater treatment, especially submerged systems such as moving biological bed reactors and integrated fixed-film systems. The latter is also used to upgrade existing activated sludge systems. Fixed-film reactors remain at the core of on-site treatment and small-scale plants as well as for tertiary treatment, especially nitrogen removal. I owe a great deal to all those inspirational researchers who have shared their works, especially the late Dr. H.A. Hawkes of Aston University, who pioneered research into the ecology of wastewater treatment and the area of applied hydrobiology. One last word about the quote in the frontispiece, which appears in all the editions and reprints of *Biology of Wastewater Treatment*. I was working late one evening hand-drawing the illustrations for the first edition. While listening to the radio, I heard Spike Milligan utter that phrase, which seemed to apply so aptly not only to myself but to all wastewater engineers and scientists. When I wrote to ask him for permission to use it, he was pleased that he belonged to a greater brotherhood and so allowed the phrase to be freely reproduced.

This book is intended for all those working and researching in the area of biofilm reactors. Although it is not exhaustive nor a design manual, it intends to give a detailed overview of the processes and some guidance to the key research that has underpinned it.

Nick Gray
May 2020

# Acknowledgments

I am very grateful to the many people who have kindly provided me information. I would especially like to thank all the publishers, organizations and companies who have given me their permission to reproduce or modify their material in the text. I am especially grateful to Drs. H.A. Hawkes, M.A. Learner, I.L. Williams and A.D. Wheatley for their help and inspiration.

My special thanks go to Amanda Yun, Senior Editor at World Scientific Publishing, for all her help and patience, and also to Gregory Lee, who has been of such great assistance in getting this book into print.

# Contents

# 1

# Introduction to Fixed-Film Reactors

## 1.1 The Process

It is clear when examining stones or other submerged objects in an enriched river or the wall of a sewer, that microorganisms will readily colonize any suitable surface, provided sufficient nutrients are present. This principle has been utilized in fixed-film or attached-growth systems where the microbial biomass is present as a thin layer that grows on the surface of a support medium. Purification is achieved when the wastewater is brought into contact with this microbial film or biofilm. Because the active biomass is largely retained within the reactor, there is generally no need to recirculate any displaced biomass back to the reactor to maintain a sufficient density of microorganisms, as is the case in the activated sludge process (Fig. 1.1). The required contact between the biofilm and the wastewater is achieved in most fixed-film reactors (FFRs) by allowing the wastewater to pass over the surface of the stationary medium on which the film has developed. However, the medium does not need to be stationary. In many reactor designs, the medium itself moves through the wastewater, ensuring maximum contact with nutrients in wastewater and the biofilm.

FFRs can be broadly grouped into seven basic designs based on the media type (i.e., random or modular), whether they are operated with a medium that is fully or partially flooded (i.e., submerged), or non-submerged, and on their integrated use with the activated

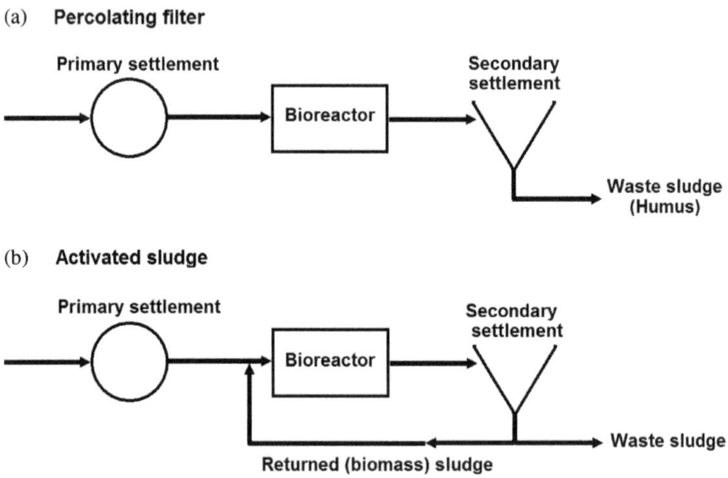


**Fig. 1.1.** Comparison of the process design of a conventional fixed-film reactor (i.e., percolating filter) with the completely mixed activated sludge process.

sludge process (Metcalf & Eddy, Inc. *et al.*, 2014). Specifically, these designs are:

- *Non-submerged fixed-film systems (aerobic)*, e.g., low-rate and high-rate percolating or trickling filters, biological air treatment systems.
- *Partially submerged fixed-film systems (aerobic)*, e.g., rotating biological contactors.
- *Sequential combined non-submerged fixed-film and activated sludge systems (aerobic)*, e.g., a high-rate filter followed by a conventional or extended aeration-activated sludge unit.
- *Submerged suspended media fixed-film systems (aerobic)*, e.g., fluidized beds and biological aerated filters, moving bed biofilm reactors.
- *Submerged fixed media fixed-film systems (aerobic)*, e.g., submerged aerated filters.
- *Fixed-film media for inclusion in activated sludge basins (aerobic)*, e.g., an integrated fixed-film activated sludge process.
- *Fixed-film systems for nitrification/denitrification (aerobic/anoxic)*.

Each process group is compared schematically in Fig. 1.2.

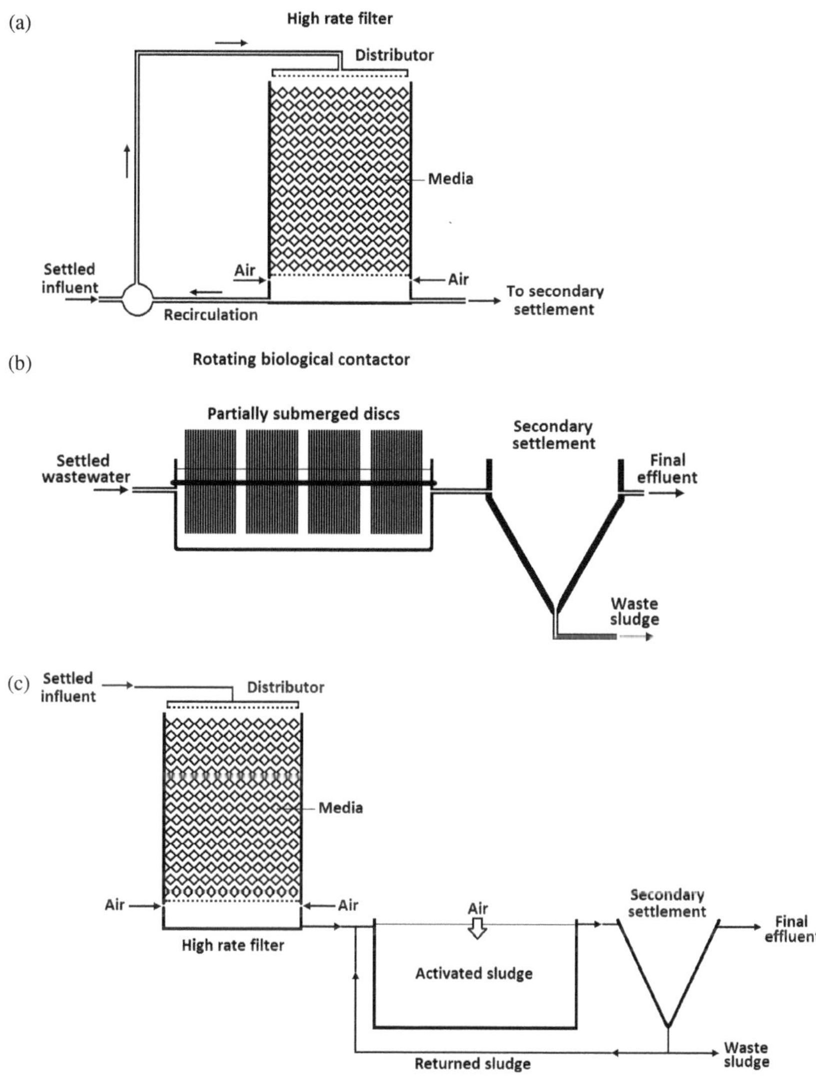


**Fig. 1.2.** Comparison of major aerobic fixed-film processes: (a) High-rate modular medium percolating filter (non-submerged system), (b) rotating biological contactor (partially submerged system), (c) high-rate filter followed by conventional activated sludge (sequential combined non-submerged and activated sludge system), (d) moving bed biofilm reactor (submerged suspended media system), (e) submerged aerated modular filter (submerged fixed media system), and (f) integrated fixed-film activated sludge process incorporating anoxic pre-denitrification.

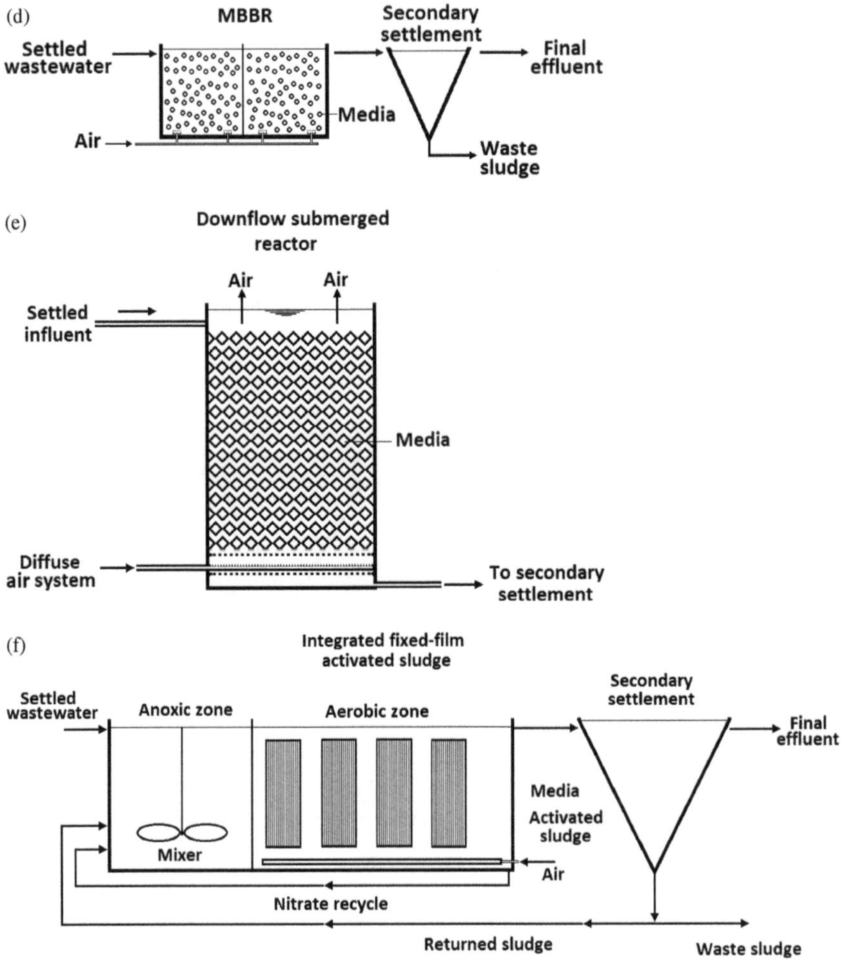


**Fig. 1.2.** (*Continued*)

All the processes rely on an inert medium or carrier on which a microbial film or biofilm develops, which is comprised primarily of heterotrophic and autotrophic bacteria. Fungi are also an important component, as is a diverse range of other microscopic organisms collectively known as grazers. In terms of their ecology, all have similar micro and mesofauna, with the non-submerged systems also having extra trophic levels comprising macroinvertebrates and other grazing animals. Biofilm formation is explored in Chap. 2 while the ecology of FFRs is examined in detail in Chap. 4, using trickling filters as a model system.

## 1.2 Application of Fixed-Film Technologies

Fixed-film reactors (FFRs) are primarily designed as secondary treatment processes although the degree of treatment can vary, as can their suitability for different wastewater types. They can be operated under aerobic, anoxic and fully anaerobic conditions, although most applications are aerobic and treat both domestic and industrial wastewaters of medium strength.

The most widely used FFR design is the percolating or trickling filter, which is considered in detail in Chap. 3. Rotating biological contactors (Chap. 5), biological-aerated flooded filters and submerged-aerated filters (Chap. 6) are similar in function to percolating filters that are used primarily for aerobic treatment of organic wastewaters, usually combined with nitrification. They are still widely employed for the treatment of settled domestic and municipal wastewaters with the final effluent discharged directly to surface waters.

However, anoxic and anaerobic filters, which also involve the development of a heterotrophic film on a suitable support medium, are also used. They are widely used to partially treat strong wastewaters with a high concentration of soluble BOD (>400 mg $L^{-1}$) and which contain little suspended solids such as those from the agricultural and food processing sectors or industrial processes. Under anaerobic conditions, the organic matter is converted to gaseous end products with little sludge being produced. In contrast, industries that are producing complex but degradable wastewaters such as those generated from pharmaceutical manufacturing plants utilize high-rate (aerobic) trickling filters (Sec. 3.3). The effluents from these systems require further treatment before disposal to receiving waters and so need to be subsequently treated by a low-rate FFR or the activated sludge process (Secs. 3.2, 5.1 and 6.1).

Nitrification of municipal wastewaters is incorporated into most standard FFR designs. However, both nitrification and denitrification can also be carried out separately by FFR units for nitrogen-rich wastewaters (Chap. 7). The most widespread use of anoxic filters is for the removal of oxidized nitrogen from purified effluents by denitrification, with the nitrogen present reduced to elemental nitrogen and released from the filter as a gas. Denitrification requires a carbon source, and in the removal of oxidized nitrogen (nitrite, nitrate) from purified effluents, this is usually supplied as methanol, although other carbon sources are also used.

Odors generated from wastewater treatment or industry are expensive and difficult to treat. Biological air treatment has largely replaced the use of physical-chemical scrubbers and can treat low-medium concentrations of many volatile organic compounds as well as the hazardous gases hydrogen sulfide and ammonia at very high gas exchange rates by biofiltration or in biotrickling filters. With a growing demand for sustainable energy, they are playing an important role in the desulfurization of biogas (Sec. 8.4).

Reed beds and sand filtration are also classified as FFRs and can be employed for enhanced BOD removal as a tertiary treatment step, with many novel developments being researched. *In situ* systems for the treatment of road run-off such as grass swales and retention ponds used in sustainable urban drainage systems are other examples of fixed-film systems.

## Reference

Metcalf & Eddy, Inc., Tchobanoglous, G., Stensel, H., Tsuchihashi, R. and Burton, F. (2014) *Wastewater Engineering: Treatment and Resource Recovery, 5th ed.* McGraw-Hill Education, New York, US.

# 2
# Biofilm Formation

## 2.1 What are Biofilms

Since the earliest times of microscopy it has been well known that microorganisms can attach and grow over the surface of wet or submerged materials. However, it is comparatively recently that these microorganisms have been discovered to have distinct characteristics that allow them to form what is commonly referred to as biofilms. While Antonie van Leeuwenhoek (1632–1723) described microorganism growth as "a film" on his teeth three hundred years ago, it was not until the mid-20th century that the importance and extent of biofilms were realized in surface waters, water supply networks and in the treatment of wastewater. The development of percolating or trickling filters, the earliest constructed fixed-film reactor (FFR), was based on the concept of filtration. Initially, the visual development of biofilms and surface microbial mats were a cause of concern for operators. It was not until the early 1900s that the role of microorganisms in the treatment of wastewater began to be realized, and this development was both encouraged and actively managed. Studies during the 1940s on surface water microbiology demonstrated that the number of active microorganisms in biofilms was significantly higher than that found in water (Zobell, 1943). This led to a new generation of FFR designs that controlled the growth rate of biofilms or incorporated media designed to provide greater specific surface areas for biofilm development.

It was in the late 1960s when Jones *et al.* (1969) examined biofilms from a trickling filter using electron microscopy and discovered that not only were they composed of a wide variety of bacterial and fungal types, they were also held together by a secreted polysaccharide that formed the scaffold or matrix, within which the microorganisms were supported. Numerous studies followed that showed biofilms to be widespread both in domestic and industrial water and wastewater systems, that they developed rapidly, and were physically stable and tenacious but highly resistant to biocides including disinfectants such as chlorine and chlorine dioxide (Characklis, 1973). A theory of how microorganisms adhere to surfaces was published in *Scientific American* in 1978, which demonstrated the advantages to microorganisms of being part of a biofilm structure, creating an important ecological niche found in all environments from acidic streams to living tissue (Costerton *et al.*, 1978). Research into biofilms in ecosystems and industrial processes run in parallel with that in medicine and dentistry, with the advent of new microscopic techniques (e.g., confocal laser scanning microscopy) giving new insights into the structure of biofilms. Biofilm development and attachment are now even better understood due to gene profiling (Donlan, 2000).

So, what is a biofilm? It is an assemblage of microorganisms (i.e., primarily bacteria, fungi and algae, which play a significant role where light is present) held together in a three-dimensional matrix of extracellular polymeric substances (EPS), which is permanently attached to a surface (i.e., is not removed by normal shear forces of the liquid phase). Biofilms that develop on living tissue (e.g., the throat) or on indwelling medical devices tend to be very simple structures dominated by a single species of coccoid bacteria (e.g., commonly staphylococcal bacteria) embedded in a very thin EPS matrix (Branda *et al.*, 2005). In contrast, water supply and wastewater biofilms are highly complex; they comprise many varied species in a thicker and denser EPS matrix. Organic debris and inert material such as clay and silt, corrosion particles, precipitation products, iron and aluminum salts can also accumulate within the matrix, being irreversibly attached to the adsorption sites within the matrix as well as on the surface of the bacteria themselves. Meiofauna (e.g., protozoa, rotifers, nematodes) are commonly associated with environmental biofilms, while macro-fauna (e.g., insect larvae) often use biofilms as a food resource and are generally classified as grazers.

> A biofilm is an assemblage of bacterial, fungi and algae on which meiofauna attach and graze. It also contains non-living organic and inorganic debris and in some cases corrosion and precipitation products, embedded in a secreted matrix of extracellular polymeric substances.

What gives biofilms their adhesion, structural stability and protects them from being destroyed by external physical or chemical factors is the EPS, which is secreted by the bacteria themselves. The EPS matrix, which comprises polysaccharides, proteins, nucleic acids and lipids, is a complex system that enables the bacteria and associated colonizers to be highly successful and resilient to multiple stressors. Recent research into the control of biofilm infections in humans has also identified the important role that extracellular DNA (eDNA) has in adhesion, formation and the stability of biofilms (Whitchurch *et al.*, 2002). Studies have shown that eDNA is predominantly produced via cell lysis induced by quorum sensing and that once released, attaches to the surface of living cells extending to form acid-base bonds to inert surfaces. eDNA is an integral part of the matrix and so by targeting the enzymatic degradation of the eDNA, the biofilm can be made less stable and more susceptible to biocides (Okshevsky *et al.*, 2014; Okshevsky and Meyer, 2015). Both the chemistry of EPS and its role in biofilm formation are complex and explored in detail by Flemming and Wingender (2010) and Lembre *et al.* (2012).

## 2.2 Formation of Biofilms

Bacterial cells are generally perceived as living in isolation from one another, but in fact, they are able to interact in many ways, including the production of microbial flocs and biofilms. Largely through medical research, we now know how individual cells can come together to form stable layers on many different surfaces, both biological and non-biological, through a process known as quorum sensing, creating complex and often highly specialized structures with a high degree of organization and co-operation between cells.

Individual bacterial cells can sense the density of other bacteria around it, and when a threshold number (or quorum) is present, they are then able to work together via a wide range of physico-chemical and biological processes to form a biofilm or other structure such as a floc. Quorum sensing appears to be a continuous process where individual bacteria produce signaling molecules or autoinducers that diffuse through the cellular membrane into the liquid phase. So, the higher the density of bacteria in an area, the higher the concentration of autoinducer molecules released, which eventually triggers a collective behavioral change in all the bacteria. The secreted autoinducers are a mixture of homoserine lactones and specific peptides that control specific repressor or activator sequences in the bacterial DNA responsible for the production of mRNA. This forms a new protein that includes among others, the genes for the formation of biofilm. Where this gene is absent an organized biofilm is not produced.

Biofilm development starts with an attachment to a solid surface. This phase is critical and occurs in two stages — reversible and irreversible attachment — with biofilms only developing after the latter stage. When clean surfaces become available, there is a certain degree of conditioning where dissolved hydrophobic organic molecules accumulate at the surface-water interface. This conditioning film, which can form very quickly (Table 2.1), is affected by the nature and surface texture of the substrate as well as the chemical nature of the liquid phase. Flow characteristics are also important, especially the shear stress, which will also have a significant influence on biofilm development in general (Loeb and Neihof, 1975). Bacteria make contact with this new surface either by passive or active movement and become attached to the surface by physico-chemical forces (e.g., electrostatic, van der Waals forces, hydrophobic effect, etc.), which are reversible. As bacteria are generally negatively charged they have a hydrophobic surface. The increasing non-polar nature of the cell and/or the surface of the medium aids attachment (Rosenberg and Kjelleberg,

**Table 2.1.** Duration of various stages in the formation of a biofilm in a fixed-film reactor.

| Time scale | Stage of biofilm development |
|---|---|
| Minutes | Conditioning of clean surface |
| Seconds/minutes | Reversible attachment of bacteria |
| Minutes | Irreversible attachment of bacteria |
| Minutes/hours | Initial cell division |
| Hours/days | Secretion of EPS and expansion of biofilm |
| Weeks | Maturation of biofilm |
| Weeks | Colonization by protozoa |
| Weeks/month | Colonization by larger meiofuana |
| Weeks/months | Possible colonization by macroinvertebrates |
| Years | Establishment of larger macroinvertebrates |

1986). Fimbriae (or attachment pili) are common in both Gram-negative and Gram-positive bacteria. They are different from flagella, which are used for locomotion (being thinner (3–8 nm) and shorter), and they are also numerous (up to 1,000 per cell). Fimbriae are important in medicine as it is through them that bacterial pathogens attach to receptors on human cells leading to infections. They are also important in adhesion to conditioned inert surfaces in the formation of biofilms (Bullitt and Makowski, 1995), probably by overcoming the electrostatic repulsion between the cell wall and medium surface. Motile bacteria are thought to be able to attach to surfaces more rapidly than non-motile strains, as they are able to overcome any repulsive forces of the medium (Korber *et al.*, 1989). However, the surface properties of individual bacterial cells such as the presence of EPS, lipo-polysaccharides, proteins and mycolic acid all increase hydrophobicity and hence adhesion potential. The next stage is the irreversible attachment of the bacteria to the surface, which is achieved by the secretion of EPS that create strong chemical bonds between the cells and the surface. The bacteria continue to produce EPS, creating a matrix in which bacteria can divide and where new bacteria are

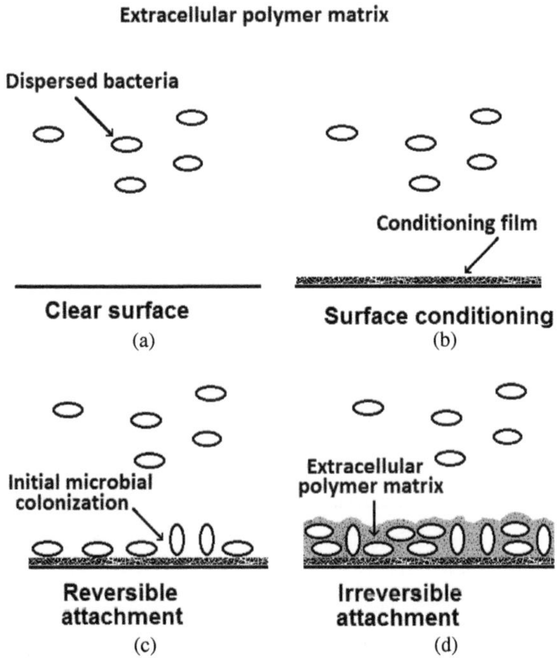


**Fig. 2.1.** Key steps in the attachment of a biofilm to an inert surface.

attracted and attached, thus creating an active biofilm (Marshall, 1985; 1992) (Fig. 2.1).

Once this initial colonization is complete the biofilm then begins to go through a series of developmental stages. Early development is often referred to as expansion, with the biofilm becoming increasingly thicker and developing a unique structure. While biofilms can still attract new bacteria (both the same and different species), most of its growth is by cell division. The structure that develops is complex, with the matrix forming different shapes such as columns, cones and mushrooms of cells with voids between each other (Costerton *et al.*, 1995; Korber *et al.*, 1995; Lewandowski, 2000; Tolker-Nielsen and Mollin, 2000; Wolfaardt *et al.*, 1994) (Fig. 2.2). This is not random but again depends on cells communicating with others through quorum sensing. The voids have an important function that allows nutrients and oxygen to penetrate from the liquid phase deep into the biofilm layer and at the same time allowing waste gases and oxidized material to be removed. In non-submerged FFR systems, the

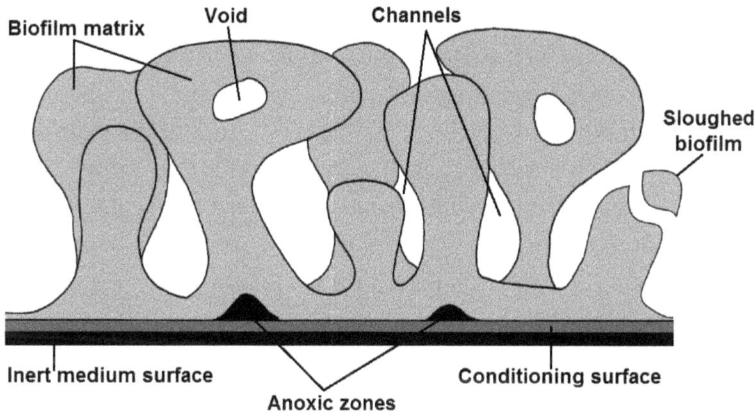


**Fig. 2.2.** Biofilms are composed of thick porous structures made from extracellular polysaccharides (EPS). Liquid can readily flow between these structures via channels. Where the biofilm is very thick, oxygen penetration may be reduced, thus creating anoxic areas. As the biofilm matures, parts can be detached by hydraulic action — a process known as sloughing.

wastewater flows over the biofilm as a thin layer, with a significant portion of the flow passing through these voids that ensure maximum contact between the nutrients in wastewater as well as allowing fine particulate matter to be physically filtered from the liquid phase and retained by the biofilm, where it is eventually oxidized. The voids are also important as they allow the film to continue to grow. In activated sludge flocs only the outer surface layer of bacteria actively reproduces, while those within the floc remain viable but with the sole function of producing extracellular enzymes that increase the rate at which adsorbed dissolved, colloidal and small particulate organic material can be broken down for absorption by cells, thereby rapidly making the adsorption sites available again. If this was not the case, the floc would continue to grow from the inside out, thus forcing the floc to break up. Similarly, active growth in biofilms is restricted to the surface with the adsorption of organic material taking place both on the EPS matrix as well as on adsorption sites on the bacterial cell walls. In this way, the EPS matrix is creating a digestive system (Flemming and Wingender, 2010) and retaining extra-cellular enzymes within the biofilm where they can release metabolized material for absorption by the bacteria. This is just one example of how the biofilm community works together to create a massive advantage over microorganisms that live as individuals, with up and down-regulation of genes controlling the release of gene products that govern the actions of individuals or layers of bacteria pivotal in community organization. More details of biofilm formation are given by Wang *et al.* (2019), who studied biofilm development in a full-scale integrated fixed-film activated sludge system, making comparisons between biofilm and activated sludge floc development.

## 2.3 Biofilm Function and Development

With time the biofilm becomes established with its thickness controlled mainly by the hydraulic stress, which is a function of the nature of the support medium and the hydraulic stress imposed by the bulk liquid on the surface of the biofilm. The next phase is maturation, where the biofilm becomes colonized by meiofauna and later by mac roinvertebrates (Table 2.1). These graze the surface of the biofilm, reducing the biomass, thereby ensuring that the voids are retained while encouraging rapid new growth. This grazing also ensures that biofilms do not become too thick, which might prevent oxygen from penetrating to the base of the matrix or result in greater frictional effects as wastewater passes, causing them to become detached. Nematodes feed within the matrix, creating new channels and voids, while fly larvae and oligochaetes, found in conventional random media percolating filters, have the same effect but also destabilize thicker biofilms at the same time, causing shedding of the outer layer — a process known as sloughing. This activity is linked closely to temperature (Sec. 3.4.5).

The permeability of biofilms is a crucial factor in treatment efficiency, which increases the contact (retention) time between the active biofilm and the wastewater. As the wastewater makes its way through the reactor, its chemical nature is altered, in particular, the organic strength as measured by the biochemical oxygen demand (BOD) or chemical oxygen demand (COD) is reduced. This substrate or nutrient gradient allows different consortia of bacteria to dominate the biofilm, with nitrifying bacteria often restricted to the lower section or end of FFRs. Higher organic loading rates at the top or beginning of the reactor result in more biofilm development composed predominantly of heterotrophic bacteria that rapidly overgrow slower growing species such as the nitrifiers.

One of the advantages of FFRs compared to other treatment processes is that they can withstand toxic or shock loadings with very little effect on overall removal efficiency. Biofilms act as a protective barrier between bacterial cells and the environment. The EPS matrix has a high adsorptive capacity that protects the bacteria within from toxic elements, including metals, trace organics, including pesticides, and even bactericides. Its importance in the treatment process cannot be stressed highly enough, as it accounts for 50 to 90% of the total carbon content of biofilms, with the amount of EPS comprising the biofilm increasing with age (Flemming *et al.*, 2000). The close proximity of toxic compounds with the bacteria results in a gradual reduction in susceptibility to potential inhibition. With time the biofilm community becomes acclimated to these compounds and will even be able to degrade them. FFRs and biofilms are used in a wide variety of industrial applications, including the treatment of toxic wastewaters such as phenols and other organics. Antibiotics are also adsorbed, which is why biofilm-based infections in humans and animals are often difficult to treat (Donlan, 2000). Antibiotic resistance is also developed and transferred to other bacteria associated with the biofilm, resulting in the possibility of multiple resistant bacteria being released into the environment (Ehlers and Bouwer, 1999).

The ability of the hydrated biofilm to act as a protective layer is used to good effect in rotating biological contactors (RBCs) where the biofilm is continuously exposed to physico-chemical stress as well as the effect of desiccation and even exposure to UV radiation, as the rotating discs on which the biofilm develops is continuously exposed to the atmosphere to allow oxygen transfer before being resubmerged for a short period in the wastewater to allow for further adsorption and rehydration.

The recent development of 454 pyrosequencing of the 16S rDNA genes has allowed the community structure of biofilms to be better understood. Intriguingly it appears that there is a high degree of organization and similarities between diverse types of biofilms. For example, biofilms have been widely employed in sand filtration in the removal of nutrients and pathogens during water treatment. Iker (2013) studied the use of biofoam, a 7-mm-thick synthetic porous medium that is able to effectively remove pathogens from water as a possible replacement for larger sand filtration in point-of-use situations. He discovered a high degree of similarity between biofilms collected from three different sampling locations, with the dominant classes being Alphaproteobacteria, Gammaproteobacteria and Betaproteobacteria, with the genus *Pseudomonas* of the class Gammaproteobacteria making up 25% of the shared sequences. Comparable results have also been published by Zhang *et al.* (2012a) who found the dominant classes to be Alphaproteobacteria and Gammaproteobacteria, while a high degree of similarity between bacterial communities in biofilms colonizing water distribution pipes was recorded by Henne *et al.* (2012). Similar studies at this level on FFR biofilms have yet to be done, although the bacterial diversity of activated sludge has been extensively examined (Ye and Zhang, 2013; Zhang *et al.*, 2012b). These early studies suggest that certain genera may be required for the successful development of a biofilm and that there may be a more organized community structure involving different key genera than previously thought. Certainly, these structures are highly organized and have sophisticated communities that will have much more to reveal to those who study them.

## References

Branda, S.S., Vik, S., Friedman, L. and Kolter, R. (2005) Biofilms: The matrix revisited. *Trends in Microbiology*, **13**, 20–26.

Bullitt, R. and Makowski, L. (1995) Structural polymorphism of bacterial adhesion pili. *Nature*, **373**, 164–167.

Characklis, W.G. (1973) Attached microbial growths: II. Frictional resistance due to microbial slimes. *Water Research*, 7, 1249–1258.

Costerton, J.W., Geesey, G.G. and Cheng, K.-J. (1978) How bacteria stick. *Scientific American*, **238**, 86–95.

Costerton, J.W., Lewandowski, Z., Caldwell, D.E., Korber, D.R. and Lappinscott, H.M. (1995) Microbial biofilms. *Annual Review of Microbiology*, **49**, 711–745.

Donlan, R.M. (2000) Role of biofilms in antimicrobial resistance. *ASAIO Journal*, **46**, S47–S52.

Ehlers, L.J. and Boiwer, E.J. (1999) RP4 plasmid transfer among species of *Pseudomonas* in a biofilm reactor. *Water Science and Technology*, **7**, 163–171.

Flemming, H.-C., Wingender, J. (2010) The biofilm matrix. *Nature Reviews. Microbiology*, **8**, 623–633.

Flemming, H.-C. and Wingender, J., Griegbe, C. and Mayer, C. (2000) Physico-chemical properties of biofilms. In *Biofilms: Recent Advances in Their Study and Control* (ed. L.V. Evans). Harwood Academic Publishers, Amsterdam, pp. 19–34.

Henne, K., Kahlisch, L., Brettar, I. and Höfle, M.G. (2012) Analysis of structure and composition of bacterial core communities in mature drinking water biofilms and bulk water of a citywide network in Germany. *Applied and Environmental Microbiology*, **78**, 3530–3538.

Iker, B.C. (2013) Application of advanced molecular techniques in applied environmental microbiology. Ph.D. Thesis. University of Arizona, Tucson, US. https://repository.arizona.edu/handle/10150/301699

Jones, C.H., Roth, I.L. and Sanders W.M. (1969) Electron microscopic study of a slime layer. *Journal of Bacteriology*, **99**, 316–325.

Korber, D.R., Lawrence, J.R., Lappin-Scott, H.M. and Costerton, J.W (1995) Growth of microorganisms on surfaces. In *Microbial Biofilms* (eds. H.M. Lappin-Scott and J.W. Costerton). Cambridge University Press, Cambridge, pp. 15–45.

Korber, D.R., Lawrence, J.R., Sutton, B. and Caldwell, D.E. (1989) Effect of laminar flow velocity on the kinetics of surface recolonization by $Mot^+$ and $Mot^-$. *Pseudomonas fluorescens. Microbial Ecology*, **18**, 1–19.

Lembre, P., Lorentz, C. and Di Martino, P. (2012) Exopolysaccharides of the biofilm matrix: A complex biophysical world. In *The Complex World of Polysaccharides* (ed. D.N. Karunaratne). IntechOpen, London, pp. 371–92,

Lewandowski, Z. (2000) Structure and function of biofilms. In *Biofilms: Recent Advances in Their Study and Control* (ed. L.V. Evans). Harwood Academic Publishers, Amsterdam, pp. 1–17.

Loeb, G.I. and Neihof, R.A. (1975) Marine conditioning films. *Advances in Chemistry*, **145**, 319–335.

Marshall, K.C. (1985) Mechanisms of bacterial adhesion at solid-liquid interfaces. In *Bacterial Adhesion: Mechanisms and Physiological Significance* (eds. D.C. Savage and M. Flecher). Plenum Press, New York, pp. 131–161.

Marshall, K.C. (1992) Biofilms: An overview of bacterial adhesion, activity, and control at surfaces. *American Society of Microbiology (ASM) News*, **58**, 202–207.

Okshevsky, M. and Meyer, R.L. (2015) The role of extracellular DNA in the establishment, maintenance and perpetuation of bacterial biofilms. *Critical Reviews in Microbiology*, **41**, (3), 341–352.

Okshevsky, M., Regina, V.R. and Meyer, R.L. (2014) Extracellular DNA as a target for biofilm control. *Current Opinion in Biotechnology*, **33**, (June), 73–80.

Rosenberg, M. and Kjelleberg, S. (1986) Hydrophobic interactions in bacterial adhesion. *Advances in Microbial Ecology*, **9**, 353–393.

Tolker-Nielsen, T. and Molin, S. (2000) Spatial organization of microbial biofilm communities. *Microbial Ecology*, **40**, 75–84.

Wang, J., Liu, Q., Wu, B., Zhao, F., Ma, S., Hu, H., Zhang, X. and Ren, H. (2019) Quorum sensing signalling distribution during the development of full-scale municipal wastewater treatment biofilms. *Science of the Total Environment*, **685**, 28–36.

Whitchurch, C.B., Tolker-Nielsen, T., Ragas. P.C. and Mattick J.S. (2002) Extracellular DNA required for bacterial biofilm formation. *Science*, **295**, 1487.

Wolfaardt, G.M., Lawrence, J.R., Roberts, R.D., Caldwell, S.J. and Caldwell, D.E. (1994) Multicellular organization in a degradative biofilm community. *Applied and Environmental Microbiol*ogy, **60**, 434–446.

Ye, L. and Zhang, T. (2013) Bacterial communities in different sections of a municipal wastewater treatment plant revealed by 16S rDNA pyrosequencing. *Applied Microbiology and Biotechnol*ogy, **97**, 2681–2690.

Zhang, M., Liu, W., Nie, X., Li, C. Gu, J. and Zhang, C. (2012a) Molecular analysis of bacterial communities in biofilms in drinking water clearwell. *Microbes and Environments*, **27**, (4), 443–448.

Zhang, T., Shao, M.-F. and Ye, L. (2012b) 454 pyrosequencing reveals bacterial diversity of activated sludge from 14 sewage treatment plants. *ISME Journal*, **6**, 1137–1147.

Zobell, C.E. (1943) The effect of solid surfaces on bacterial activity. *Journal of Bacteriology* **40**, 547–548.

# 3

# Percolating (Trickling) Filters

## 3.1 Introduction

The design and function of percolating filters have been described by numerous researchers (Pike, 1978; Bruce and Hawkes, 1983; Parker, 1999; Boltz *et al.*, 2010). In its simplest form, the reactor consists of a filter medium, a bed of graded hard material that is approximately 2 m in depth. The medium has interstices or voids that allow both air and applied wastewater to reach all parts of the reactor. The air is supplied to all parts of the reactor by a passive ventilating system, while a distributor regulates the volume and frequency of application of wastewater (influent) over the surface (Fig. 3.1). The medium provides the necessary base for the attachment of non-motile microorganisms, principally bacteria and fungi, which form the biofilm (Chap. 2). Motile organisms, both micro and macroscopic, live within the biofilm layer or in the shelter of the interstices, feeding on and controlling the accumulated biofilm. The action of this grazing fauna controls biofilm growth, thus preventing heavy accumulations blocking the interstices (Hawkes, 1963), which would cause ponding and anaerobic conditions within the reactor. The accumulation of the biofilm follows a seasonal pattern where it becomes thicker during the winter months. The action of the grazing fauna loosens and breaks down the film, resulting in a large removal of biomass each spring, which is known as sloughing. The nutrients in the wastewater promote the growth of microorganisms that comprise the biofilm, thus as the wastewater

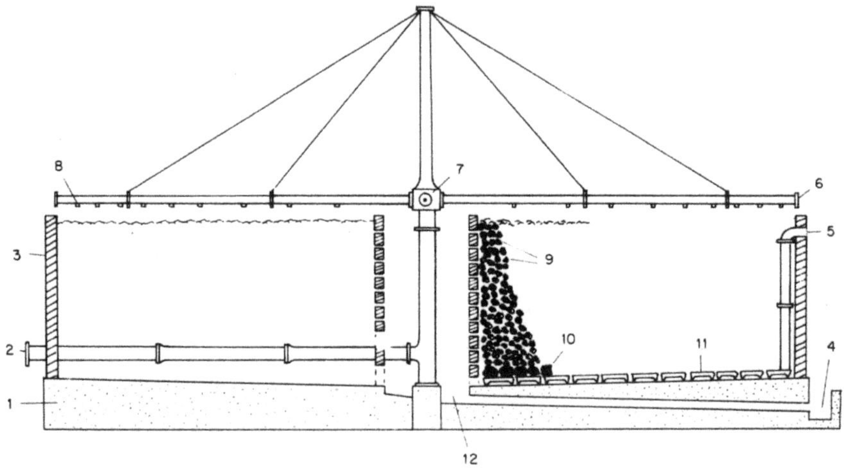


**Fig. 3.1.** Basic constructional features of a conventional percolating filter: 1. Foundation floor, 2. Feed pipe, 3. Retaining wall, 4. Effluent channel, 5. Ventilation pipe, 6. Distributor arm, 7. Rotary seal, 8. Jets/nozzles, 9. Main bed of medium, 10. Base layer of the larger medium, 11. Drainage tiles, 12. Central well for effluent collection (Bruce and Hawkes, 1983). Reproduced with permission of Elsevier Publishing.

percolates downward over the surface of the biofilm-covered medium, biological oxidation and conversion take place.

Percolating filters are the second-most widely used secondary treatment process in Europe, being equally common in other temperate regions, including the United States (US), where percolating filters (also known as trickling filters) are employed at over 3,700 separate municipal wastewater treatment plants. Although biological filtration is historically the first non-land-based wastewater treatment process developed, it still has certain advantages over the more common biological secondary treatment process — activated sludge. Percolating filters require virtually no skilled maintenance or close control. In energy terms, they are more economical than the activated sludge process and more versatile in responding to changes in hydraulic loading and the characteristics of wastewater (Hawkes, 1963). These filters are more tolerant of continual or shock discharges of certain pollutants compared with activated sludge (Cook and Herning, 1978), including toxic industrial wastes containing heavy metal ions, phenols, cyanides, sulfides and formaldehyde. They are widely used for both total and partial treatment of a wide range of industrial wastewaters (Calley *et al.*, 1977), with a growing use as stand-alone nitrification or denitrification units (Pike, 1978; Barnard and Steichen, 2006; Lu *et al.*, 2014; Zhu *et al.*, 2016). Their major disadvantage is capital cost because they become uneconomic when serving populations of greater than 50,000. This is due not only to higher capital costs (as compared to activated sludge systems) but also to the larger area of land they occupy, which is often at a premium in urban areas (Table 3.1). For this reason, the activated sludge process predominates at very large domestic and municipal treatment works. Even as late as 1970, the proportions of the populations of England and Wales served by these two biological oxidation processes were about the same (Institute of Water Pollution Control, 1972). Since that time, however, there has been a steady decline in the use of percolating filters for domestic treatment, although they remain popular for treating industrial wastewaters. Where fixed-film reactors have proved popular is for the treatment of domestic wastewaters for small populations of <2,000 PE, and at sites requiring systems with a small physical footprint, where rotating biological contactors or submerged aerated filter designs are ideal. Reed beds are increasingly being used for secondary and tertiary treatment for small rural plants serving <1,000 PE. So while standard percolating filtration is less popular, it remains a reliable and sustainable option that will become more important as wastewater treatment plants try to reduce their carbon footprints.

The term "percolating filter" is still widely used, especially in the United Kingdom (UK), although there are many derivations of the name such as trickling filter, biological filter, bacteria bed, and percolator. However, the term "filter" often confuses, as the process is essentially biological, even though there is some physical removal of fine solids. The confusion arises because when the process was first developed in the late 19th century, it was thought that purification was brought about by physical filtration, as in certain types of sand filter, and that percolating filtration was developed from such physical filter processes. Experiments were carried out at the Lawrence Experimental

**Table 3.1.** Comparison between the activated sludge and percolating filter process.

| | Activated sludge | Percolating filtration |
|---|---|---|
| Capital cost | Low | High |
| Area of land | Low: Advantageous where land availability is restricted or expensive | Large: 10 times more area required |
| Operating cost | High | Low |
| Influence of weather | Works well in wet weather, slightly worse in dry weather, less affected by low winter temperatures | Works well in summer but possible ponding in winter |
| Technical control | High: The microbial activity can be closely controlled; requires skilled and continuous operation | Little possible except process modifications. Does not require continuous or skilled operation |
| Nature of wastewater | Sensitive to toxic shocks, changes in loading, and trade wastewaters; leads to bulking problems | Strong wastewaters satisfactory, able to withstand changes in loading and toxic discharges |
| Hydrostatic head | Small: Low pumping requirement, suitable for site where available hydraulic head is limited | Large: Site must provide natural hydraulic head otherwise pumping is required |
| Nuisance | Low odor and no fly problems. Noise may be a problem both in urban and rural areas | Moderate odor and severe fly problem in summer possible. Quiet in operation |
| Final effluent quality | Poor nitrification but low in suspended solids except when bulking | Highly nitrified, relatively high suspended solids |
| Secondary sludge | Large volume, high water content, difficult to dewater, less stablized | Small volume, less water, highly stablized |
| Energy requirement | High: Required for aeration, mixing, and maintaining sludge flocs in suspension and for recycling sludge | Low: Natural ventilation, gravitational flow |
| Synthetic detergents | Possible foaming, especially with diffusers | Little form |
| Robustness | Not very robust, high degree of maintenance on motors, not possible to operate without power supply | Very sturdy, low degree of maintenance, possible to operate without power |

Station of the Massachusetts State Board of Health using small filter beds to test the efficiency of various soils and aggregates, e.g., different grades of sand and gravel, to physically remove solids from wastewater. They found that effective purification of settled sewage could be achieved using coarse gravel (19–25 mm grading) and at very high rates of application (0.2–0.5 $m^3m^{-2}d^{-1}$), so it was not always necessary to use sand filtration or land treatment, both of which required much lower hydraulic loading rates to be as effective (Mills, 1890). Although it was quickly realized that this new type of treatment involved a degree of biological oxidation, it was not until the process was well established that the physical role of filtration was found to be minor in comparison to the biological function, by which time the terms "trickling" and "percolating filtration" had become established. The impact on sewage treatment practice was immediate because the process only required one-tenth of the area that existing land treatment (i.e., ridge and furrow sewage farming) employed, and as the beds were specially constructed, treatment became independent of the suitability of the soil for the first time. The first percolating filters were constructed in Salford (England) in 1892. This first installation was designed by Corbett (1902), and it was here that many of the practical problems of design and construction were solved. The first municipal trickling filter installed in the US was in Atlanta, Georgia, and was commissioned in 1903 (Stanbridge, 1976; Bruce and Hawkes, 1983).

## 3.2 Design and Operation

### 3.2.1 *Design*

The design of percolating filters has changed very little since they were first introduced, and with a working life of 80 to 100 years, many of the earliest reactors are only now reaching the end of their useful lives. However, while the filter medium does eventually deteriorate and require replacement, process failure is more often due to organic or hydraulic overloading rather than structural or mechanical failure. The major factors that need to be taken into account in the design and operation of percolating filters are: medium type, in particular, specific surface area and voidage, depth, area of the bed, and both organic and hydraulic loading rates.

The depth of the medium (the filter bed) is dependent on the structural stability of the reactor walls, the degree of treatment required and the need to sustain aerobic conditions throughout the bed. When all types of percolating filtration are considered, then there is an approximate range in medium depth of between 0.9 and 15.0 m. The standard (low-rate) percolating filter, commonly used in the UK, is usually about 1.8 m deep and rarely less than 1.5 m or greater than 2.5 m. Filter beds in the rest of Europe tend to be deeper with filters in Germany, for example, between 3 to 4 m in depth. Low-rate filters in the US are of a similar depth to those in the UK, at between 1.5 to 2.0 m, although shallow beds (<1 m) can be employed for high-rate filtration. High-rate filters using modular plastic medium are used to treat industrial wastewaters. The prefabricated medium comes in large units which are very light so that they can be stacked, like building blocks, into a tall tower that is essentially free-standing. These are referred to as biotowers and are constructed in lightweight prefabricated materials as high as 12–15 m. Smaller units are contained in simple block or brick structures.

In terms of treatment efficiency, the majority of biochemical oxygen demand (BOD) and suspended solids removal occur within the top 750 mm of the filter bed, which is also about the minimum depth when using a conventional mineral medium to avoid short-circuiting of wastewater through the bed. Apart from constructional costs, the major limiting factor in the selection of the depth of filter beds is the loss of hydraulic head; the greater the depth, the greater the loss of head occurring. So, unless the plant is built on a slope, pumping will be required. From experience, a depth of 1.8 m, when using a conventional medium, is a good compromise between treatment efficiency, loss of hydraulic head, and cost. The influence of depth on performance is considered in Sec. 3.4.4.

There is a wide variety in the configuration of percolating filter systems, although to a large extent, this is dependent on the site and loading rates. Individual filters are usually rectangular or circular in plan. Rectangular filters are normally used for large populations because they are more compact and are cheaper to construct as they share retaining walls. The maximum size of such filters rarely exceeds 10 m in width and 100 m in length. Rectangular designs are used for very small populations (<40 people), which is more compatible with the mechanical distribution system available. This comprises of a tipping trough that discharges the wastewater intermittently into a fixed set of distribution channels that are laid across the filter. Circular plan filters are generally preferred as they allow greater control over the frequency of wastewater application. The maximum size of these filters is limited by the maximum diameter of the rotating distributor available, which is generally 40 m.

Percolating filters are built on a concrete base covered with drainage tiles that provide a raised floor on which the medium rests (Fig. 3.1). This provides an unrestricted area of under-drainage, which takes the final effluent away. It is important to ensure that the under-drainage area is also deep enough to allow the passage of air from ventilation pipes lining the base of the filter bed. The ventilation pipes are connected to the atmosphere so that there is a free flow of air passively through the whole filter, providing the necessary aeration. Natural convection causes aeration because of the difference between the air temperature and the temperature within the bed and can be in either direction depending on the temperature difference. The concrete base is sloped to at least 1:100, normally 1:50, to ensure a minimum flow of 0.6 m $s^{-1}$, which prevents settlement and the accumulation of solids washed from the medium. Drainage tiles can be replaced by a series of half-round field drains laid out in a herringbone pattern in rectangular filters or arranged to radiate out from the center of circular-shaped filters to a central drainage channel or well (Fig. 3.2) (CIWEM, 2000). As is the case with the tiles, the drains must be of sufficient size to allow for adequate ventilation.

Filter bed walls must be able to support a considerable weight of the medium, attached film and water (Table 3.2). The retained material exerts lateral pressure on the walls, which increases toward the base, and it is thus cheaper and safer to construct filter beds partially or completely buried in the ground. This allows blocks or a single layer of bricks to be used for the retaining wall, and it being below the

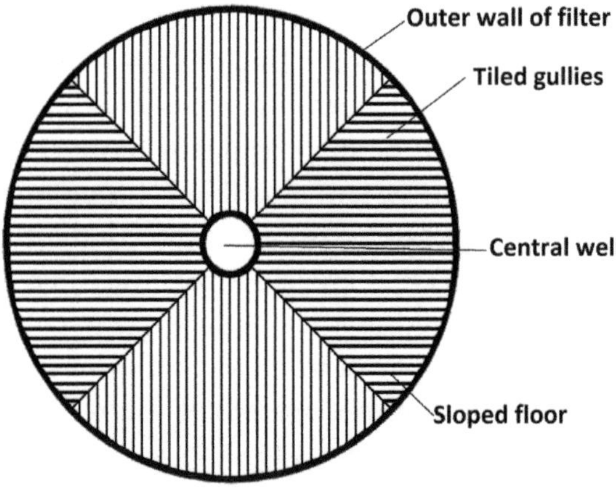


**Fig. 3.2.** Typical arrangement of filter floor tiles draining to a central well.

**Table 3.2.** Comparison of the bulk density of a mineral medium and a random plastic medium of the same nominal grade (50 mm).

| | Dry weight (kg m$^{-3}$) | Total bulk density due to film accumulation (kg$^3$ m$^{-3}$) | | Weight when totally saturated with water (kg m$^{-3}$) |
|---|---|---|---|---|
| | | Minimum film | Maximum film | |
| Flocor RC | 97.84 | 133.73[a]<br>140.68[b] | 290.59[a]<br>354.77[b] | 1004.84 |
| 50 mm slag | 886.31 | 1014.55[a]<br>1050.01[b] | 1277.32[a]<br>1180.33[b] | 1401.31 |

[a]Low-rate loading, [b]high-rate loading.

ground surface ensures that maximum thermal insulation is provided by the surrounding soil. Filter walls are normally under ring tension with the forces involved increasing with the diameter of the filter (Haywood and Cross, 1976). When constructed entirely above ground, the retaining walls must be able to take the maximum bulk density of the filter medium contained within the bed. In this case, the walls are constructed out of reinforced concrete, thick brickwork, or prefabricated concrete sections bolted together. When precast sections are used, the joints do not need to be watertight as the lateral spread of effluent in the bed is limited under low to moderate hydraulic loadings. At higher loadings, some seepage of wastewater may occur at the joints. In the early years, some small filter beds were built free-standing. This was done by inclining the sides sufficiently so that the media formed a stabilized bank, which was then covered using a larger grade of aggregate. Drystone walling was also popular, although both systems were liable to the walls collapsing and surface water overflowing down the sides. Modular plastic media exerts a negligible amount of lateral pressure on the retaining walls, and therefore lightweight cladding, usually coated steel panels, or a single layer concrete block housing are adequate. This is not the case with random plastic filter media, as is seen when the maximum bulk density exerted on the retaining walls can be almost equal to that of a mineral medium if ponding occurs (Table 3.2) (Gray, 1983a).

The main function of the distribution system is to apply the wastewater to the top of the filter bed as evenly as possible. Distribution systems can be categorized as either static or moving. Static distributions are a network of pipes or open troughs that are permanently fixed in position above the surface of the medium. The wastewater is discharged through fixed outlets that are usually spray nozzles (where pipes are used) or via V-notches from open channels. However, the uniformity of dosing and the completeness of wetting the surface of the medium is not as good as obtained when using moving distributors, and so static distribution systems are restricted to very small percolating filters. High-rate filters using modular media are often square or rectangular in plan (Sec. 3.1.3), and because it is important to have even distribution to utilize all the potential surface area of the medium, static spray nozzles are frequently used. There are two types of moving distributors: rotary distributors that are used for circular filters and reciprocating distributors that are employed on rectangular beds. Rotary distributors have either 2 or 4 radial sparge pipes that are pivoted in the center of the filter on a rotating column (Figs. 3.1 and 3.3). The outlet holes are spaced along the arm to ensure a uniform loading, with the holes becoming more closely spaced further away from the center. The outlets must be large enough to ensure that blockages from solids carried over from the settlement tank or accumulated grease rarely occur.

**Fig. 3.3.** Mechanically driven surface distribution arm. The wastewater flows from nozzles without splash plates onto the surface of the percolating filter filled with mineral medium.

Once blocked, the area of the bed fed by that outlet hole will be without any influent until the blockage is cleared. For this reason, simple holes, with splash plates situated below to ensure maximum dispersion of the wastewater, are preferred to jet or spray nozzles. The distributor arms can be driven by water wheels, electrically powered wheels that drive the distributor arm around or in smaller beds by jet reaction as the water leaves the outlet. Whereas jet reaction does not require a power supply and is mechanically simple, it requires a liquid head of 0.5 m as well as an extremely low friction bearing to support very well balanced distributor arms. The speed of rotation varies according to the hydraulic loading. Such systems are affected by changes in wind direction, which can result in the distributor stopping altogether and causing localized overloading of the bed with a subsequent reduction in effluent quality (IWEM, 2000).

Reciprocating distributors are associated only with filters that are rectangular in plan and are only common in the UK. The walls of the bed serve as a track to convey as well as support the distributor as it moves back and forth along the bed. Wastewater is fed into the moving sparge pipe from a longitudinal supply channel by syphonic action (IWEM, 2000). The moving distributor is propelled either by a water wheel mechanism or an electric motor using flexible cables. Because the distributor cannot travel more than 10 m $min^{-1}$, it can take a considerable time to travel from one end to the other. This can result in long intervals between successive dosing of wastewater, especially at the far ends of the bed, thus longer beds may employ several distributors, each operating over a particular area (Stanbridge, 1972).

### 3.2.2 *Loading*

For conventional single-pass filtration where settled sewage passes through a single percolating filter bed before having the solids (humus) removed in a secondary sedimentation (humus) tank, loading can be categorized as either low- or high-rate. Hydraulic loading is expressed in cubic meters of settled wastewater per cubic meter of filter medium per day ($m^3m^{-3}d^{-1}$) with loadings of $<3$ $m^3m^{-3}d^{-1}$ classed as low-rate and $>3$ $m^3m^{-3}d^{-1}$ as high-rate filtration. However, there is no precise demarcation between low and high-rate filtration, so these figures are only guidelines. Hydraulic loading can also be expressed as a surface loading rate, also referred to as the irrigation rate. This is measured in cubic meters of wastewater per square meter of superficial bed area per day ($m^3m^{-2}d^{-1}$). The hydraulic loading rate depends solely on the physical characteristics of the medium and the degree of biofilm accumulation within the bed. The upper hydraulic limit is determined by the onset of surface flooding because the wastewater is being applied faster than it can be percolated through the bed. Where this occurs, then the interstices of the medium become flooded, and the access of air to the interior of the bed is prevented. The lower limit is known as the minimum wetting rate, which is the loading rate at which the medium is kept sufficiently moist to prevent drying and for the microbial biofilm to remain active (Winkler, 1981).

Organic loading is measured as the BOD per cubic meter of filter medium per day (kg BOD $m^{-3}d^{-1}$) with low-rate filtration classed as filters receiving loads of <0.6 kg BOD $m^{-3}d^{-1}$ and at a high rate of >0.6 kg BOD $m^{-3}d^{-1}$. To produce the old Royal Commission standard effluent of 20 mg $L^{-1}$ BOD and 30 mg $L^{-1}$ suspended solids (20:30) after settlement with a high degree of nitrification, filters treating domestic sewage should receive an organic loading of between 0.07–0.10 kg BOD $m^{-3}d^{-1}$ and a hydraulic loading between 0.12–0.60 $m^3m^{-3}d^{-1}$. Generally, increases in organic loading in excess of 0.10 kg BOD $m^{-3}d^{-1}$ will result in heavier biofilm growths, which may result in ponding. To produce more efficient percolating filters able to operate at much higher loadings, a number of modifications of the basic process have been adopted. By using a larger mineral medium, for example, greater loads can be applied to filters without the risk of ponding. Such high rate filtration will produce a 20:30 effluent with increased hydraulic loading of up to 1.8 $m^3m^{-3}d^{-1}$, but with little or no nitrification (Institution of Public Health Engineers, 1978). If a less stabilized effluent is required, such as roughing treatment for strong industrial wastes, then loadings of up to 12 $m^3m^{-3}d^{-1}$ with organic loads of up to 1.8 kg BOD $m^{-3}d^{-1}$ will give a 60–70% BOD removal. Treatment at such high rates is facilitated by using a modular plastic medium (Fig. 3.4) in tall towers, or a random plastic medium (90% voids) (Fig. 3.5) in place of the usual stone medium (40% voids) (Fig. 3.5), thus reducing the risk of ponding (Fig. 3.6). Modifications of the percolating filter process are dealt with in Sec. 3.3.

If the amount of biological biofilm growing per unit area of the medium remains more or less constant, then the organic loading corresponds to the sludge loading or the food to microorganism ratio (f/m ratio) as used in the activated sludge process. This is difficult to measure directly as biofilm accumulation is rarely even throughout the filter bed and is certainly more abundant in the upper areas of the bed, where food is a non-limiting factor. Also, a considerable proportion of the biomass will be non-active, comprised of inert macro-invertebrate fragments, which increases toward the base of the filter. In theory, using sludge loading in a percolating filter offers a degree of control on par with that of the activated sludge process, but in practice, accurate estimations of biofilm accumulation can only be made for high-rate filters, especially for those using a modular plastic

**Fig. 3.4.** Surface of a biotower housing modular plastic medium Flocor E™. Splash plates are used to ensure even distribution of influent wastewater over the surface of the medium, ensuring maximum utilization of the interior surface area.

medium where seasonal variation in biofilm development and the problem of inert solids being retained in small interstices have been largely eliminated. Another important factor is that none of the medium in a high-rate filter should be either food or oxygen-limited, thus ensuring an even growth of biofilm throughout the bed. Estimates of biofilm accumulation have been made. For example, the active depth of biofilm is known to be about 0.2 mm, which gives an average weight of active biofilm per square meter of a medium of around 0.2 kg. If it is assumed that the water content of biofilm is about 97%, then each square meter of a medium's surface area is supporting 6 g (DW) of active microbial biomass. The exact amount in the filter will depend on the actual surface area of the medium, but if this is assumed to be 100 $m^2m^{-3}$, then the active microbial biomass will be 0.6 kg $m^{-3}$. Thus, at an organic loading of 0.1 kg BOD $m^{-3}d^{-1}$ the sludge loading will be about 0.2 kg BOD $kg^{-1}d^{-1}$. For high-rate systems, assuming the same surface area for the medium, the sludge

(a)

(b)

**Fig. 3.5.** The surface biofilm development on (a) blast furnace slag, and (b) random plastic filter media (Flocor RC™).

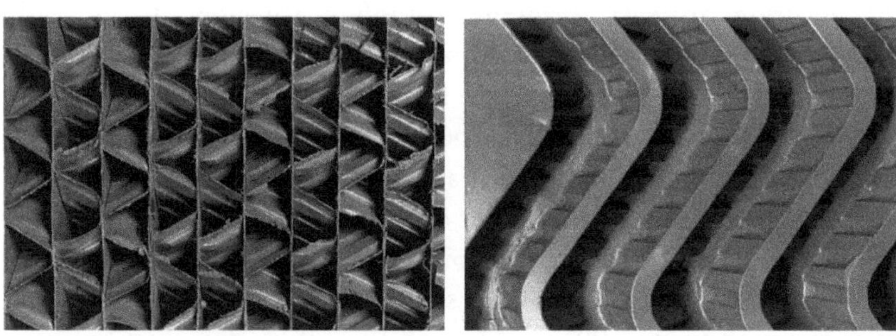

(a) (b)

**Fig. 3.6.** Configuration of the modular plastic medium Flocor E™. (a) The top surface over which the wastewater is applied via spray nozzles to give even distribution, and (b) the crossflow nature of the medium prevents free fall of the wastewater and provides a high surface area on which the biofilm develops.

loading will be approximately 1 kg BOD $kg^{-1}d^{-1}$ at an organic loading of 0.6 kg BOD $m^{-3}d^{-1}$. Therefore, there would appear to be an agreement between the term low- and high-rate as used in both the percolating filtration and the activated sludge processes (Winkler, 1981).

### 3.2.3 *Filter Media*

The function of the medium in a percolating filter is to provide an extensive surface to support the biological film and the associated fauna necessary for the purification of settled wastewaters. It must also allow sufficient ventilation for the process to operate aerobically and ensure maximum contact between the active biofilm and the waste liquid.

*Mineral media*

The important features of mineral media were summarized in the original British Standard of 1948 (BSI, 1948). It stated that a mineral filter medium should be selected for "high surface area consistent with adequate voidage, satisfactory grading (within specified size limits), durability, roughness of texture, satisfactory shape characteristics and low cost". Thompson (1925) showed that the material selected had to be mechanically and chemically stable if the medium was not to degrade filling the voids with smaller pieces of aggregate broken off from the parent medium. Whereas Levine *et al.* (1936) compared different filter media and found that performance was directly related to the surface area of the medium. After the publication of the 1948 British Standard, it became the practice for engineers to choose a medium that was predominantly cubic. This view was based on the supposition that a filter medium containing a high proportion of flattish pieces would have an undesirably low voidage. In many textbooks, this remains the view. The more irregular a material is for a given nominal size, the greater the surface area. Therefore, blast furnace slag or clinker is preferred to gravel or pebbles, although it is more costly. For the material of a given shape or uniform grade, the voidage is independent of size. However, the important factor is the actual dimensions of the void spaces (the interstices), as these will determine whether or not a given medium will clog (i.e., pond) during operation or will allow adequate ventilation. Excessive accumulation of biofilm also reduces the effectiveness of the surface area. The rejection of flaky material (i.e., particles with one excessively thin dimension) by the British Standard 1438:1948 was proven unjustified by Schroepfer (1951). When examining the effect of particle shape on voidage and surface area, he found that material that was more cubical possessed a lower voidage than the flaky material, and he concluded that an increase in the angularity of particles, of which flaky particles are an extreme example, increased both the voidage and surface area. Compaction will reduce the available voidage by between 5–7% (Moncrieff, 1953). In a review of the relationship of particle shape and voidage, Bruce (1968) supported the earlier findings of Schropfer — flaky media possess a higher voidage and surface area than regular media of the same sieve size. However, the average volume of flaky particles is smaller, and this controls the actual size of the voids, although the use of flaky material of a large grading would compensate for this. These findings led to a relaxation of the "index of flakiness" and the withdrawal of the "index of elongation" in the revised British Standards on Percolating Filter Medium (BSI, 1971; 1980).

Numerous comparative investigations were conducted in the UK into the ideal characteristics of filter media followed the publication of the British Standard in 1948, in particular, those carried out at Minworth near Birmingham (Hawkes and Jenkins, 1955; 1958) and at Stevenage (Wilkinson, 1958; Truesdale *et al.*, 1962). These investigations showed that the smaller media consistently produced better quality effluents and that a media with a rough surface gave a marginally better performance than smooth surfaced materials such as gravel. Truesdale *et al.* (1962) found that although a small grade, rough-textured medium was extremely efficient in treating large organic loads of settled sewage during the summer (0.18 kg BOD $m^{-3}d^{-1}$), it suffered from excessive biofilm accumulation during the colder winter months. This resulted in ponding and the eventual clogging of the filter. Experience has shown that to achieve maximum efficiency throughout the year, a 50 mm mineral medium with a rough surface provides the best compromise between large surface area and the provision of large voidage and that this will produce a high-quality effluent in a conventionally operated European plant (Hawkes, 1963).

Learner (1975a), in a survey of percolating filters, found that granite (26%), clinker (24%) and blast furnace slag (23%) were the most frequently used media in the early 1970s. While the pitted nature of the clinker and blast-furnace slag makes them superior to other mineral media, the choice of filter medium does not only depend on its suitability but also availability and cost. As Learner points out, the majority of filters using clinker as a medium were constructed before 1956, and since then, clinker has become more expensive and extremely difficult to obtain. Another interesting aspect of the survey was that the majority of the filters sampled in Scotland used granite as a medium. It may not have been the most suitable medium, but it was the most readily available in the area. A similar situation exists in Ireland, where neither blast furnace slag nor clinker is available (Earle and Gray, 1987). For this reason, the majority of percolating filters use the locally available stone, which is normally basalt or granite. Learner (1975a) found that the most common range of media grades used was 13 to 51 mm based on a particle size analysis (i.e., 13–25 mm 21%, 25–38 mm 39% and 38–51 mm 29%). Bruce (1969) states in his review of percolating filters that the most common grades of media used in the UK are the nominal sizes of 38–51 mm, which is in contrast to the study by Learner. As previously stated, 50 mm mineral media provides a good compromise between surface area and voidage and has been shown to produce good quality effluents under low-rate conditions (Hawkes and Jenkins, 1955; 1958). Since Learner's survey, percolating filters have been largely replaced by more modern systems, including fixed-film reactors (FFRs) such as rotating biological contactors (RBCs) (Chap. 5) and submerged fixed biofilm reactors (Chap. 6). When media replacement is required in existing percolating filters, plastic alternatives are generally used.

*Physical nature of media*

The shape of the media is defined by using a descriptive classification based on visual inspection. They can be classed as being flat, angular, irregular, rounded or spherical. The different shapes are beautifully compared in a series of photographs in the British Standard 1438 (1971). The voidage increases with an increase in the proportion of flaky media, whereas angular-shaped media have a higher voidage than rounded media. Typically, basalt and granite media are either flat or angular (50–54% voidage), blast-furnace slag is irregular (50%), and gravel is rounded or spherical (38–40%). The voidage can be determined directly using a cylindrical metal vessel with spouted outlets at two levels and having a volume of at least 25 liters between the two outlets. The voidage is calculated by first filling the vessel full of media and then filling it with water. It may be necessary to leave the medium submerged in water for 24 hours to ensure that absorption by the aggregate is complete before topping up the vessel with water. The highest outlet is opened and allowed to drain fully and the volume of water filling the interstices of the known volume of medium between the two spouts is determined by measuring the volume of liquid that drains from the lower outlet. The results are expressed as a percentage. The volume and overall diameter of the cylinder must be as large as practicable to minimize the wall effects (Bruce, 1968). The British Standard 1438 was originally published in 1948 and revised in 1971 and 2004. The latest version of this standard should be consulted when testing potential mineral media for percolating filters.

Grading is normally done by the supplier of the medium who issues a certificate showing the nominal grade. It is important to obtain a medium that has a fairly uniform particle size, otherwise, the voidage, and more importantly, the size of the voids will be seriously reduced if there is a wide discrepancy between the sizes. When the aggregate is crushed, considerable dust and small fines are produced that must be removed by washing with high-pressure hoses before the medium is placed in the filter bed. If this is not done, localized blockages of the voids deep within the filter will eventually occur. It is impossible unless the media is manufactured, e.g., plastic media, for the medium to be comprised of particles of a single size. For this reason, the medium is graded by determining the percentage by weight passing each sieve, and the nominal size of the medium is then calculated using the limits set by the British Standard 1438 (Table 3.3). Small quantities of over- or under-sized medium do not affect the

**Table 3.3.** Grading limits for percolating filter media in the British Standard 1438 (BSI, 1971).

| **BS410 square hole perforated plate test sieves (mm)** | **Nominal sized (mm)** | | | | | |
|---|---|---|---|---|---|---|
| | **63** | **50** | **40** | **38** | **20** | **14** |
| | **% by weight passing** | | | | | |
| 75 | 100 | — | — | — | — | — |
| 63 | 85–100 | 100 | — | — | — | — |
| 50 | 0–35 | 85–100 | 100 | — | — | — |
| 37.5 | 0–5 | 0–30 | 85–100 | 100 | — | — |
| 28 | — | 0–5 | 0–40 | 85–100 | 100 | — |
| 20 | — | — | 0–5 | 0–40 | 85–100 | 100 |
| 14 | — | — | — | 7 | 0–40 | 85–100 |
| 10 | — | — | — | — | 0–7 | 0–40 |
| 6.3 | — | — | — | — | — | 0–7 |

voidage significantly, thus for a nominal size of 50 mm, for example, up to 15% of the medium by weight can be >50 mm and up to 30% can be <37.5 mm. Particle size analysis is done using square hole sieves, according to the British Standard 410:1986, using a minimum sample of 100 kg of medium (BSI, 1986). In practice, grading proves more difficult as the percentage passing each sieve often does not conform exactly to the limit set in Table 3.3. Earle (1986) gives an example where the slag medium was crushed to a nominal size of 50 mm but was found on a sieve analysis to have 100% (by weight) passing the 63.0 mm sieve; 70.5%, 50 mm; 15.1%, 37.5 mm; and 0%, the 28.0 mm sieve. This medium does not conform to the limits set for 50 mm as 29.5% was retained in the 50 mm sieve, so crushing was not done adequately. However, it would not appear serious enough to abandon the media, especially after it has been delivered, so what nominal grade is it? It does not fit the limits set for 63 mm either, so the nominal grade of the medium lies somewhere between these two values. Table 3.3 makes it clear that the bulk of the medium of the 50 mm nominal size should be between 37.5–50.0 mm in size. In this case, 55.4% was within this particular range. As up to 30% of the medium is allowed to pass the 37.5 mm sieve, Table 3.3 indicates that 55% within 50–37.5 mm is legitimate for a medium of 50 mm nominal grade, even though it specifies limits of 85–100%. Therefore, the sample tested appears to fit best within this category rather than the 63 mm nominal size grade. Interestingly, as the filter media ages, its particle-size distribution expands as pieces of the medium are weathered and fractured (Table 3.4). For this reason, the use of an average or median size for filter media has been proposed (Earle, 1986). The

**Table 3.4.** Problems of calculating the nominal size of the mineral filter medium using Table 3.3 when the medium has undergone weathering within the filter. Examples are given for Irish percolating filters and the nominal size compared with the mean size (mm) (Earle and Gray, 1987). Reproduced with permission of Tigroney Press.

| | Bulk percentages passing the sieve grading (nm) | | | | | |
|---|---|---|---|---|---|---|
| **Site** | **16** | **20** | **28** | **37.5** | **50** | **63** |
| Ballygar | 5 | 28 | 26 | 8 | 25 | 8 |
| Enniskerry | 0 | 0 | 1 | 13 | 55 | 31 |
| Gort (i) | 5 | 10 | 40 | 40 | 5 | 0 |
| Gort (ii) | 10 | 10 | 34 | 30 | 3 | 14 |
| Kildare (i) | 9 | 19 | 59 | 13 | 0 | 0 |
| Kildare (ii) | 9 | 18 | 59 | 14 | 0 | 0 |
| Killincarring | 41 | 36 | 23 | 0 | 0 | 0 |
| Leighlinbridge (i) | 2 | 4 | 54 | 38 | 3 | 0 |
| Leighlinbridge (ii) | 8 | 9 | 45 | 34 | 3 | 0 |
| Loughrea | 11 | 14 | 24 | 29 | 22 | 0 |
| Mountbellew | 0 | 0 | 2 | 41 | 56 | 0 |
| Newcastle | 1 | 2 | 11 | 41 | 46 | 0 |
| Rathmore | 8 | 19 | 45 | 23 | 6 | 0 |
| Roundwood | 1 | 2 | 8 | 24 | 46 | 19 |
| Saggart (i) | 3 | 3 | 9 | 17 | 45 | 22 |
| Saggart (ii) | 0 | 2 | 2 | 38 | 45 | 13 |

**Table 3.4.** (*Continued*)

| Site | Voidage (%) | Nominal size (mm) | Mean size (mm) |
|---|---|---|---|
| Ballygar | 42 | 37.5 | 25 |
| Enniskerry | 49 | 50 | 44 |
| Gort | 44–45 | 37.5 | 27 |
| Kildare | 42–43 | 28 | 23 |
| Kilincarring | 44 | 20 | 16 |
| Leighlinbridge | 46–47 | 37.5 | 28 |
| Loughrea | 44 | 37.5 | 29 |
| Mountbellow | 45 | 37.5 | 40 |
| Newcastle | 49 | 37.5 | 37 |
| Rathmore | 44 | 37.5 | 24 |
| Roundwood | 43 | 50 | 42 |
| Saggart | 45–48 | 50 | 40 |

British Standard 410 was withdrawn in 2000 and replaced by BS ISO 3310:2013 (BSI ISO, 2013). Methods of test sieving are given in a separate standard BS1796-1:1989 or ISO 259-1:1988 (BSI, 1989).

In selecting the medium, a compromise must be reached between the conflicting requirements of a high specific area, which requires a fine grade, and large voids, which are obtained by using a coarse grade. The surface area ($m^2m^{-3}$) is affected by surface texture and particle shape, although the nominal size of the medium is the major influence (Table 3.5). A rough, porous medium such as clinker or blast-furnace slag can have a surface area of up to 17% greater than smooth-surface media, such as gravel, of the same nominal size. However, as the biofilm develops, many of the small holes in rough-textured medium become filled with organic debris, fauna, or biofilm, thus, in practice, the surface area will tend to approach that for smooth-medium types. The numerous holes of various sizes and the irregular areas on the rough-textured medium provide a much greater variety of niches for both the microorganisms and the larger grazers than on smooth-medium types. The rougher texture also prevents the

**Table 3.5.** Values for the specific surface area ($m^2m^{-3}$) of three different types of the filter medium in relation to nominal size and grading. The range for graded material represents the maximum oversize or undersize material permitted by the British Standard 1438.

| | Type of medium | | | | | |
|---|---|---|---|---|---|---|
| Nominal maximum size of medium (mm) | Granite | | Blast-furnace slag | | Crushed gravel | |
| | Single-size | Range of graded material | Single-size | Range of graded material | Single-size | Range of graded material |
| 25 | 194 | 185–237 | 208 | 200–246 | 176 | 169–208 |
| 37.5 | 135 | 129–149 | 146 | 140–163 | 125 | 120–140 |
| 50 | 97 | 94–111 | 104 | 101–118 | 89 | 86–101 |
| 63 | 75.5 | 73–85 | 81 | 79–90 | 69 | 67–77 |

biofilm from being so readily sloughed off and allows increased weights of biomass to accumulate within the filter.

It is not normally necessary to measure the surface area of medium directly, as the type and the grade being used will most likely conform to one of the three media in Table 3.5. Where necessary, the surface area can be determined by the paint dipping technique. Using a known bulk volume of medium, each piece is coated with free-flowing paint using a standard dipping procedure (Schroepfer, 1951). The weight of the paint adhering to the medium is calculated by weighing the paint container before and after dipping. The increase in weight per unit area is calculated by dipping test blocks of a known surface area into the same paint. These test blocks can be of wood or aggregate, provided that the paint adheres to them, and the surface area can be accurately calculated. Problems are encountered in obtaining an even distribution of paint over all the pieces of the medium even after several dippings. Truesdale *et al.* (1962) noted that filter media has different absorptive capacities, and in practice, absorption of paint varies between individual pieces of the same medium. The problem of uneven coating and paint collecting and thickening at the base is overcome by painting individual pieces by hand and then calculating the increase of weight for each piece of the medium. This procedure is more time consuming than paint dipping but ensures that the coat of paint is even and that all the pores are covered. The surface area is then calculated by comparing the increase in weight of the media with the mean increase in weight obtained with the test blocks of directly measurable surface area. To overcome the varying absorptive capacity of the media the pieces of medium will have to be painted a number of times. Experiments have shown that the paint retained on test blocks varied considerably on the first painting, but this variability became less with each subsequent coating of paint. Three coats of paint appear sufficient for the most porous media. Alternative potential methods of estimating the surface area of media surfaces are available (Dudley *et al.*, 2001) and include dipping media in a PVC resin that can be subsequently peeled off in sections and scanned so that the surface area can be measured using image analysis software (McWhinnie *et al.*, 2004).

The media must be strong enough to take the weight of the medium and accumulated biofilm above it, be chemically non-reactive, and be durable so that the media does not need to be replaced for many years. The media will undergo weathering within the bed because of constant irrigation with wastewater and surface microbial activity. On the surface of the bed, the media will also be subject to normal weathering processes such as freezing and heating. Once the media begins to disintegrate, the voids will become gradually clogged with fragmented particles. This is a particular problem with organic media. In general, most sedimentary rocks are unsuitable for use as percolating filter media. The media should remain completely stable for at least 50 years. Suitability can be assessed by the "sodium sulfate soundness test". Test pieces of the medium are immersed in saturated solutions of sodium sulfate for a number of days before being dried at 105°C. This cycle is repeated 20 times, and the material is then examined for signs of possible defects or loss of weight (BSI, 1971; 2004). Any medium that passes such a severe test will certainly survive 50 years in a percolating filter. However, some media that fail this test may also be suitable. The media should be chemically non-reactive with all types of wastewater that are normally treated in percolating filters. Problems of leaching from media do occur, especially from metals, and it has been suggested that this may interfere with some sensitive species, successfully colonizing the media, such as the nitrifying bacteria. Sulfur compounds can be leached from certain slag media and may lead to damage of the concrete base of the filter under certain circumstances (Lavender, 1970). Although under normal operating conditions, excess sulfate will be diluted by the wastewater. Limestone is widely used both in Ireland and the UK, and can be used successfully, provided that the wastewater is either neutral or alkaline. Soft limestones and chalk are eventually broken down even in neutral wastewater, and where the wastewater includes the drinking water supplied from a catchment with acidic geology, dissolution will occur and the limestone medium will literally be dissolved away. Limestone should not be used to treat industrial wastewaters for this reason. Organic media, such as peat, are occasionally used and are discussed later in this section.

### *Plastic media*

This medium has been freely available in Europe since the end of the 1950s and is now usually made out of PVC or PET. It has a very high voidage (normally >90%), wide interstitial spaces, high specific surface area, and is only about 10% of the weight of the mineral medium when in operation. This allows tall filters to be built using prefabricated and light-weight structures, which is significantly cheaper than building conventional mineral filters. However, plastic filter media must be robust and strong enough to withstand the weight of 3–12 m of media and the accumulated biofilm (Table 3.2). This strength is supplied not by the material itself, but in the form it has been prefabricated. The other major advantage over mineral medium is that blockages due to accumulated biofilm are very unlikely due to the high voidage, and so high organic loads can be applied. The wide interstices allow sufficient air to reach all areas of the filter, whereas oxygen availability can be limiting in mineral filters. The specific surface area of the plastic medium varies between 83–330 $m^2m^{-3}$, although it appears that there is very little practical difference in surface area between plastic and mineral media (Tables 3.5 and 3.6).

**Table 3.6.** Comparison of the physical properties of widely used random and modular plastic filter medium with 50 mm blast furnace slag.

| Medium | Composition | Specific surface area ($m^2$ $m^{-3}$) | Voidage (%) |
|---|---|---|---|
| **Random** | | | |
| *Norton* | | | |
| Actifil 90E | Polypropylene | 101 | 95 |
| Actifil 50E | Polypropylene | 124 | 92 |
| Actifil 75 | Polypropylene | 160 | 92 |
| *MT* | | | |
| Filterpak 1127 | Polypropylene | 120 | 93 |
| Filterpak 1130 | Polypropylene | 190 | 93 |
| Mini ring (2) | Polypropylene | 118 | 93 |
| Mini ring (3) | Polypropylene | 79 | 94 |
| *ICI* | | | |
| Floor RC | PVC | 330 | 91 |
| **Modular** | | | |
| Cloisonyle | PVC | 220 | 94 |
| *ICI* | | | |
| Flocor E | PVC | 90 | 95 |
| Flocor M | PVC | 135 | 95 |
| *Munsters* | | | |
| Plasdek B 2760 | PVC | 100 | 95 |
| Plasdek B 19060 | PVC | 140 | 95 |
| Plasdek B 12060 | PVC | 230 | 95 |
| **Mineral** | | | |
| Blast furnace slag (50 mm) | | 143 | 51 |

Apart from the requirements already mentioned, the main requirements for plastic media are similar although, not identical as for mineral media. The advantage of manufacturing a filter medium is that no compromise is required. Generally, the main features of the ideal plastic media are: an ability to treat high hydraulic and organic loads, an open structure in order to avoid restricting biofilm development or subsequent blockages due to excessive biofilm growth, wide interstices to permit thorough aeration, chemical inertness, durability, physical strength to support sufficient media and associated biofilm to ensure adequate treatment, light bulk density so that structural costs can be reduced, the configuration of the media that ensures uniform distribution and minimizes channeling, and, finally, it must be cost-effective (Chipperfield, 1967; Hemmings, 1979; Harrison and Daigger, 1987).

There are two main categories of plastic media — modular (ordered) and random — with a wide variety of commercially available designs within each category that have been widely evaluated operationally (Bruce and Merkens, 1973; Forster, 1977; Besselievre and Schwartz, 1976; Winkler and Thomas, 1978; Porter and Smith, 1979; Rowlands, 1979).

Two main designs of a modular medium are generally used — vertical sheets, ribbed or corrugated, and vertical corrugated tube media. The former is the most widely used design of modular media and is made up of sheets of corrugated ribbed plastic assembled to form a rectangular module with a large number of open zigzag channels (Fig. 3.6). The modules can be prefabricated on site but are normally delivered already made up into blocks. The media is simply stacked into the filter bed, and the modules can be cut to size to fit the configuration of the bed. The individual sheets of plastic, which comprise the modules, are orientated vertically so that both surfaces of the channel are exposed to the wastewater. The design of the zigzag channel is such that wastewater is unable to free fall through the medium without coming into contact with the biofilm (i.e., cross-flow). The media is available in a range of spacings that allow for the extra accumulation of the film. The smooth texture and vertical placement of the media discourage excessive build-up of the film, and the biofilm regularly sloughs because of its weight (Pearson, 1965; Eden *et al.*, 1964). A widely used media of this type is Flocor E™, which has a dry bulk density of 40 kg $m^{-3}$ and increases to up to 300 kg $m^{-3}$ as the biofilm develops. Because of such a large increase in weight when in operation, beds over a 3-m depth will need intermediate support (Hemmings 1979; Porter and Smith 1979). It has a specific surface area of 90 $m^2m^{-3}$, comparable to blast-furnace slag of a nominal size of 63 mm, which is also used for high-rate treatment (Tables 3.5 and 3.6).

Cloisonyle™ is a frequently used vertical tube medium consisting of a block of closely assembled vertical tubes between 4 to 6 m in length (i.e., vertical flow), which extends the whole depth of the filter bed. Each 80 mm diameter tube is subdivided into 14 smaller concentric tubules, each about 15 mm in diameter. The medium has a high voidage (94%) and a specific surface area (220 $m^2m^{-3}$) (Table 3.6). However, successful operation is due entirely to an adequate distribution of wastewater, because as the tubes are not interconnected, any tubule not receiving a wastewater supply will remain dry throughout. Also, as the surfaces are entirely vertical, there is a chance that wastewater can fall for some distance within a tubule without coming into contact with the attached film. It was quickly realized that the modular media were of limited value in the complete treatment of domestic wastewater. However, modular media are extremely effective as roughing filters, removing large weights of BOD per unit volume of media at relatively low levels of efficiency in terms of BOD concentration, i.e., 50–80% reduction. Pearse and Jarvis (2011) evaluated the performance of 10 crossflow modular filters after 10 years of continuous operation and found them to be operating well. However, two common operational problems were identified. Firstly, the problem of the loss of nitrifying capability due to temperature fluctuations, and secondly, overgrazing of the biofilm by large numbers of aquatic snails (Sec. 4.4.3). This overgrazing also affected nitrification even when the temperature variation was minimized by controlling the airflow using ventilation control (Sec. 3.4.6).

Even with careful design, the short contact time between the influent wastewater and the film, and the possibility of the free fall of wastewater, prevents the modular forms of plastic media from producing well-purified effluents. To increase the contact time but retain the high voidage and high surface area per unit volume of modular media, various new random plastic media were developed. These were shaped in such a way that the free fall of the wastewater experienced in some modular designs was prevented. Its random nature increased the contact time with the biofilm while maintaining the high voidage, thus reducing the problem of biofilm accumulation and lack of ventilation experienced with mineral media (Ramsden, 1972). Plastic random media developed from the use of pall and Raschig rings, which are widely used in the chemical industry, and originally made from metal and unglazed ceramic, respectively. It was the standard Raschig rings, ceramic tubes that are 25 mm wide and 25 mm in length, that were first used as an alternative medium in percolating filters for standard mineral medium. Today, random packed media are made up of individual pieces of PVC or polypropylene molded into specific shapes in the same nominal sizes used for mineral media such as 37.5, 50 and 65 mm, although some designs can be obtained in sizes up to 90 mm in diameter. The media are generally rings or short tubes with the surface area increased by external or internal fins and ribs (Fig. 3.7). The media are packed randomly into the filter bed in the same way as mineral media. The advantages of random plastic over mineral media include large voidage, higher specific surface areas (Table 3.6), and extremely low bulk density (Table 3.2), which permits easy handling. Its lightweight nature allows it to be housed in prefabricated beds such as PVC sheeting attached to a mild steel framework, glass reinforced plastic (GRP), or in bricks. The higher

**Fig. 3.7.** Configuration of the random plastic medium Flocor RC™.

voidage allows the influent to be continuously loaded using fixed sprays, although conventional distribution systems can also be used.

Plastic media have been successfully used to treat effluent wastewater from a wide range of industries. However, whereas modular media are used for high-rate filtration, random plastic media are essentially used for low-rate applications. On the other hand, random plastic does not appear any more efficient than comparable mineral media at conventional low-rate loadings (<0.1 kg BOD $m^{-3}d^{-1}$) and may even be less efficient in terms of nitrification. Also, a comparable plastic low-rate filter will be more expensive than a mineral system. Random plastic media are generally used for the primary biological treatment of strong wastewaters before their treatment by conventional secondary treatment (Anon, 1979; Hemming, 1979). The performance of different media is compared in Sec. 3.4.1.

### *Alternative media*

Biofilm will develop on a substrate when exposed to constant wetting by wastewater and when air is available. Except for metals and glazed ceramics, almost any surface can be utilized, including living plant material.

The significant cost of specifically manufactured plastic media and the problems of the weight and handling costs of relatively cheaper mineral media means that there is a constant search for alternative new cost-effective filter media. Submerged filters, biofilters used for odor control and those used for denitrification use specific medium types, including sand, glass particles, sponges, light-weight porous fired ceramics, and shells. String or strip media has largely replaced the use of random plastic media in tall towers for industrial applications, where the development of the fungus *Subbaromyces splendens* has caused problems of internal ponding, leading to tower instability and occasional collapse (Gray, 2010).

There has also been increasing interest in the use of waste or recycled materials as potential media for FFRs of all types. These include macerated plastic bottles, bottle tops, and macerated types. For example, Mondal and Warith (2008) successfully treated leachate from a domestic landfill site by using a trickling filter packed with waste tires that had been shredded. The material was classified as either tire chips or tire crumbs and had a high specific surface area that supported a thick (2 mm) biomass layer. Interestingly, the material behaved very similarly to random plastic media with regular sloughing of the biomass. Removal efficiencies were between 81–96%, 76–90%, and 15–68% for BOD, COD and ammonia, respectively. Comparable results have also been obtained by Naz *et al.* (2014), who tested the medium under both aerobic and anaerobic conditions. Apart from stone wool cubes, there have been few mineral-based media tested in the past two decades (Kirjanova *et al.*, 2011).

While brushwood and wood pallets have been widely used as media in the construction of temporary filters for many decades, recent research has focused on the use of other organic materials. These include wood chips, bark or peat (Lens *et al.*, 1994), as well as waste agricultural process materials such as maize husks. Wood chip filters have been widely used to treat a wide range of wastewaters, including dairy parlor run-off (Murnane *et al.,* 2016). While wood chips have proven disappointing for the aerobic treatment of domestic wastewaters, they act as a successful medium in denitrifying (anoxic) bioreactors, acting as a support for the biofilm and providing the necessary organic substrate required for denitrification (Lopez-Ponnada *et al.,* 2017).

Peat has been widely used as a filtration medium providing physical filtration and strong adsorption capabilities, but more importantly it supports a microbial ecosystem similar to that found in other FFRs. Peat is partially fossilized plant material and is complex in its structure with lignin and cellulose as its major constituents. The lignin, in particular, bears polar functional groups such as alcohols, aldehydes, ketones, acids, phenolic hydroxides and esters that are involved in chemical bonding. This complex nature makes peat an excellent adsorbent for both metals and organic molecules including volatile organic compounds, pesticides and oils (Clark and Pitt, 2009; Kalmykova *et al.,* 2009; Balan *et al.,* 2010; Tang *et al.*, 2010; Alvarez-Hornos *et al.,* 2011). However, the adsorptive capacity of the medium is limited, so the removal rate of recalcitrant organic and inert compounds will eventually decline as the adsorption sites become permanently utilized. Peat is also highly absorbent and is able to retain significant quantities of water, which ensures long retention time within the filters enabling enhanced microbial oxidation, in particular. Like other filter media, the peat fibers support a diverse range of both heterotrophic and autotrophic bacteria, meiofauna (Protozoa, Rotifera) and higher forms, in particular, oligochaetes.

Peat filter media is available commercially in different forms, either as peat moss or more commonly as fibrous peat, which is normally used for water and wastewater treatment. Milled peat is rarely used as it readily ponds. Fibrous peat has been widely used as a filter medium in the treatment of a variety of wastewaters including sewage (Koiv *et al.*, 2009), combined sewer overflows (Tao *et al.*, 2009), storm-water and road run-off (Clark and Pitt, 2009), and landfill leachate (Koiv *et al.*, 2009). The process is also employed in the on-site treatment of septic tank effluent (Martin and Anderson, 2007) and has also been used for the treatment of septic tank effluent from two schools (Kennedy and Van Geel, 2010; Garzon-Zuniga and Buelna, 2011).

Although peat filters have been studied for use as a tertiary treatment method for secondary treated wastewaters for over 40 years (Surakka and Kamppi, 1971; Farnham, 1974) most studies on the treatment of settled sewage or septic tank wastewaters have been largely laboratory-based (Viraraghavan and Rana, 1991; Corley, *et al.*, 2006), which has resulted in peat filtration being underutilized for domestic wastewater treatment, especially in areas where inadequate subsurface percolation exists. This has been more so for small group systems. Comparing vertical to horizontal flow peat filters, Koiv *et al.* (2009) recorded significantly higher BOD, chemical oxygen demand (COD) and nitrogen removal in vertical systems. They also observed that in both systems, the removal of total phosphorus declined after six months of operation.

Laboratory column studies have all shown peat to be a potentially excellent filtration media for the treatment of septic tank effluent (Viraraghavan and Rana, 1991; Corley *et al.*, 2006), and it has been successfully used in low-cost, low-maintenance wastewater treatment systems for single houses, such as the Puroflow™ System. However, its ability to treat the wastewater from larger multiple house developments has not been widely reported in the literature. Gray and Coady (2015) studied the performance of single-pass vertically loaded fibrous peat biofilters in treating septic tank wastewater from a single household (PE 6) and a small rural housing estate (PE 100). Both systems achieved a >90% BOD removal and consistently low suspended solids of <11 mg $L^{-1}$. The pathogen removal was generally >95%, although it was inconsistent. Nitrification was linked to temperature with seasonal removal rates (50–98%) with no denitrification, while phosphorus removal was generally poor. The filters responded poorly to sudden increases in hydraulic loading, causing a reduction in BOD removal in particular.

The use of waste organic media has been extensively researched. For example, Aslam *et al.* (2017) used waste agricultural material as a filter medium for the partial treatment of domestic wastewater before reuse as irrigation water. Cotton sticks, 1.3 m in length and 1.25–2.50 cm in diameter were placed vertically in stainless steel cylindrical pilot percolating filters, and domestic wastewater was distributed evenly over the surface. Removal efficiencies of 69–78% BOD, 65–80% COD and 38–56% for suspended solids were achieved over a four-month experimental period at surface hydraulic loading rates of between 3.7 and 10.1 $m^3$ $m^{-2}$ $h^{-1}$. No degradation of the sticks was recorded, and it was concluded that they could be used as a cheap locally sourced filter medium. In a similar experiment, processed maize cobs were tested as a medium for trickling filters by Ali *et al.* (2016). Over six months at a surface loading of 113 $m^3m^{-2}d^{-1}$, removal efficiencies of 79% BOD, 75% COD and 44 suspended solids were reported. The media supported a good layer of biofilm, although treatment was incomplete. Maize cobs were also used as effective biosorbents for metals and organic chemicals (Bhatnagar and Sillanpaa, 2010) as well as dyes (Saroj *et al.*, 2015). There has also been extensive research done on the use of the peeled dehydrated fruits of *Luffa cyllindriica*, the sponge gourd, a member of the cucumber family that is used to produce the familiar fibrous and highly porous luffa. The plant is especially common in Brazil where it has been successfully employed at the laboratory scale to support heterotrophic and autotrophic biofilms in both percolating and submerged attached growth systems (Sousa *et al.*, 2008; Agra, 2009; Ruiz-Marin *et al.*, 2009; Vianna *et al.*, 2012). Although BOD, suspended solids and nitrification have been similar to stone media, the eventual decomposition of the media and its physical support characteristics when wet are obvious drawbacks in a conventional trickling filter design. Like maize cobs, luffa has also been successfully used as a biosorbent for the removal of divalent metals from solutions (Laidani *et al.*, 2010; Oboh *et al.*, 2011) as well as dyes (Oliveira *et al.*, 2007). Agave plants are used in Mexico to produce the spirit mescal. The waste fibers from the process plants have been successfully tested by Vigueras-Cortés *et al.* (2013) as a medium in mechanically aerated biofilters treating municipal wastewater. The experiment ran continuously for 17 months with 92.0%, 79.7% and 91.9% removals of BOD, COD and suspended solids, respectively, with high helminth (98.0%) and fecal coliforms (99.9%) removals also. Similarly, date palm fibers have also be found to be a suitable medium (Riahi *et al.*, 2009)

### 3.2.4 *Purification and Biofilm Development*

The microbial biofilm takes 3–4 weeks to become established during the summer and up to 2 months in the winter. Only then would the filter have reached its optimum purification capacity, although nitrification will take longer. Kirjanova *et al.* (2011) recorded steady-state carbonaceous removal in their pilot scale filters using stonewall media in just two weeks, but that nitrification required three months to become established and a further month to achieve steady-state conditions. The biofilm is a complex community of bacteria, fungi, protozoa, and other mesofauna, plus a wide diversity of macroinvertebrates, such as enchytraeids and lumbricid worms, dipteran fly larvae, and a host of other groups that actively graze the biofilm (Curds and Hawkes, 1975) (Sec. 4.4). Unlike many industrial wastewaters, it is not necessary to seed domestic wastewaters treated by percolating filters because all the necessary microorganisms are present in the sewage itself, with the dipteran grazers flying on to the filter and colonizing it (Ulug and Uncuncu, 1992; Kirjanora *et al.*, 2011). Other macrofauna come from the household waste stream, primarily from washing vegetables or are washed into the sewer via surface water run-off. The debris in roadside gully pots is rich in macrofauna, especially Annelida, which are readily displaced during heavy precipitation resulting in scouring of the lighter organic debris from the pot into the sewer. The microorganisms quickly form a biofilm over the available medium. However, the biofilm only develops on surfaces that receive a constant supply of nutrients. Therefore, the effectiveness of the media to redistribute the wastewater within the filter, in order to prevent channeling and to promote maximum wetting of the medium, is an important factor affecting performance. It is possible to seed FFRs to speed up bacterial and grazer colonization by either adding good quality activated sludge to the reactor (Ray and Murthy, 1998; Shavelle *et al.*, 2008) or adding biofilm from a humus tank (Zhu and Chen, 2002). Bioaugmentation using a commercially available cultured seed manufactured specifically for FFRs is also possible (Zhou *et al.*, 2008).

The organic matter in the wastewater is degraded aerobically by heterotrophic microorganisms that dominate the film. The biofilm has a spongy structure, rather similar to activated sludge flocs, which is made more porous by the feeding activities of the grazing fauna that are continually burrowing through the biofilm (Sec. 2.2). The wastewater passes over the surface of the biofilm, and to some extent through it, although this depends on the biofilm's thickness and hydraulic loading (Fig. 3.8). In low-rate filters, a large proportion of the wastewater may be flowing through the biofilm matrix at any one time, and it is the physical straining action of this matrix that allows such systems to produce extremely clear effluents. Another advantage is that the greater the proportion of wastewater that flows through the biofilm, the greater the microorganism-wastewater contact time, i.e., the retention time or hydraulic retention time (HRT). The higher the hydraulic loading, the greater the proportion of the wastewater passing over the surface of the film, which results in a shorter HRT and a slightly inferior final effluent. Whether the liquid forms a separate layer passing over the surface of the biofilm or is making its way slowly

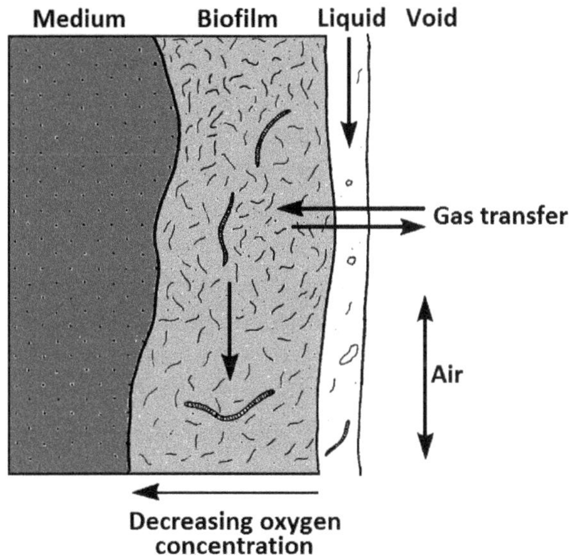


**Fig. 3.8.** Diagrammatic representation of the structure of the biofilm with internal grazers within the biofilm.

through the biofilm matrix, the first stage of purification is the adsorption of organic nutrients onto the film. Fine particles are flocculated by extracellular polymers secreted by the microorganisms and adsorbed on to the surface of the film, where they are broken down (together with the organic nutrients that they are physically trapped with) by extracellular enzymes secreted by heterotrophic bacteria and fungi. The soluble nutrients in the wastewater and those produced from this extracellular enzymatic activity are directly absorbed by the microorganisms comprising the biofilm and then synthesized. Oxygen diffuses from the air in the interstices, first into the liquid, and then into the film. Conversely, carbon dioxide and the end products of aerobic metabolism diffuse in the other direction. The thickness of the biofilm is critical, as the oxygen can only diffuse for a certain distance through the biofilm before being utilized, leaving the deeper areas of the biofilm either anoxic or anaerobic. The depth to which oxygen will penetrate depends on a number of factors, such as the composition of the film, its density and the rate of respiration within the biofilm itself, which has been estimated as being between 0.06–2.00 mm (Schulze, 1957). Tomlinson and Snaddon (1966) experimentally calculated this critical depth to be 0.2 mm for a predominately bacterial film. However, where fungal mycelium is present, oxygen diffusion is enhanced by movement within the hyphae, increasing the critical depth to 4 mm. Only the surface layer of the biofilm is efficient in terms of oxidation, and a thin layer of biofilm is required for efficient purification. Tomlinson and Snaddon estimated that the optimum thickness in terms of performance efficiency was only 0.15 mm. This means that it is the total surface area of active biofilm that is important and not the total biomass of the biofilm (Sec. 2.3).

The biofilm accumulates and subsequently thickens within the filter during operation. This is due to an increase in microbial biomass from the synthesis of the waste and to the accumulation of particulate material by flocculation and physical entrainment where the accumulation rate exceeds the rate of solubilization and assimilation by the microorganisms. Where the wastewater is largely soluble, such as food processing wastewaters, most of the biofilm increase will be due to microbial growth. Whereas with domestic wastewater that has been poorly settled, the accumulation of solids may account for the major portion of the biofilm accumulation. Once the biofilm exceeds the critical thickness, an anoxic, and subsequently, an anaerobic, environment is established below the aerobic zone. As the thickness continues to increase, most of the soluble nutrients will have been utilized before they can reach the lower microorganisms, thus forcing them into an endogenous phase of growth. This has the effect of destabilizing the biofilm because the lower microorganisms lose their ability to hold on to the surface of the medium, which results in portions of the biofilm becoming detached and washed away in the wastewater flow, a process known as sloughing. Although thick biofilm growths do not reduce the efficiency of the filter, excessive growths can reduce the volume of the interstices, reducing ventilation and even blocking them completely, preventing the movement of wastewater. Severe clogging of the interstices is known as ponding and is normally associated with the surface of the filter when whole areas of the surface may become flooded.

The accumulation of biofilm within a filter bed follows a seasonal pattern — being low in the summer due to high metabolic and grazing rates but high in the winter when the growth rate of the microorganisms is reduced, as is the activity of the grazing fauna. As the temperature increases in spring, there is a discernible sloughing of the biofilm that has accumulated over the winter months (Fig. 3.9). Wheatley and Williams (1976) found that no single factor directly influenced biofilm accumulation within their experimental filters but suggested that it was controlled by the interaction of a number of factors, namely ambient temperature, organic load, the distribution system, the microbial characteristics of the film, and the activity of grazers. Temperature was shown to be an important factor in biofilm accumulation by Hawkes and Shephard (1971), who demonstrated

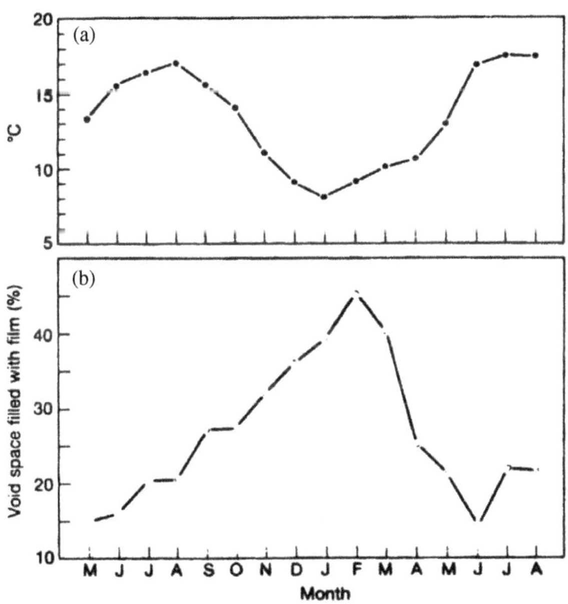


**Fig. 3.9.** Seasonal variation in the temperature of the sewage applied to a percolating filter and in the percentage void space of the filter filled with biofilm and water. (a) Monthly average temperature of applied sewage, and (b) monthly average proportion of voids filled with biofilm. Quantity of biofilm and water measured by the neutron scattering technique (Bruce *et al.*, 1967).

that the rate of biofilm accumulation increased rapidly below 10°C. In a comparative experiment using laboratory filters with and without macroinvertebrate grazing fauna at 5°C and 20°C, respectively, they examined the effects of grazers and temperature in biofilm accumulation (Shephard and Hawkes, 1976). They found that at higher temperatures a greater proportion of the BOD removed by adsorption was oxidized, and therefore less biomass accumulated. The rate of oxidation decreased as the temperature fell, although the rate of adsorption remained unaltered. Therefore, there was a gradual increase in biomass accumulation at lower temperatures, which eventually resulted in the filters becoming clogged.

Seasonal variation in biofilm accumulation in the US in percolating filters is thought to be due to this differential microbial activity at different temperatures (Cooke and Hirsche, 1958). However, such variations in the UK have been primarily attributed to the activity of grazing fauna, with the biofilm accumulating during the winter months when both population densities and the activity of grazers are at their lowest densities (Tomlinson, 1946b; Hawkes, 1957). It appears that the grazing fauna plays a significant role in reducing the overall biofilm biomass by directly feeding on the film, converting it into fecal pellets and body biomass that are flushed from the filter in the effluent (Hawkes and Shephard, 1972). Spring sloughing of the accumulated biofilm is pre-empted by the increased activity of the grazers, which loosen the thick biofilm from the medium. Once the biofilm has become detached it strips accumulated biofilm from other regions of the bed as it is washed out. While grazers suppress biofilm accumulation and maintain the thickness of the biofilm for a long period after sloughing, it is the temperature that primarily controls overall biofilm accumulation. Whereas hydraulic loading in low-rate filters is of little significance compared to the action of macroinvertebrate grazers in controlling film, as the hydraulic loading increases (as is the case after modifications to the process such as recirculation of double filtration) physical scouring of the biofilm by the wastewater becomes increasingly important. In high-rate filters, especially those employing modular plastic media, the high hydraulic loading controls the biofilm development by scouring the biofilm from the smooth-surfaced media as it reaches critical thickness. In such filters, sloughing tends to occur on a more regular basis rather than seasonally. Biofilm development is considered further in Sec. 3.4.5.

The organic matter in the settled sewage applied to a percolating filter is present as fine particulate and colloidal solids, or in solution. The biofilm transforms this material by biophysical processes (flocculation and adsorption) and biochemical processes (bio-oxidation and synthesis) to produce a variety of end products. Some of these are soluble or gaseous end products such as nutrient salts or carbon dioxide from oxidation processes, or soluble products from the lysis of the microorganisms comprising the film. What remains are solids that require separation from the final effluent. These solids are of three types: flocculated solids, detached fragments of the accumulated biofilm and the grazing fauna, and fragments of their bodies and feces. These solids are collectively known as humus, and like all secondary solids they require further treatment after settlement. The mode of operation of percolating filters influences the nature of the humus sludge produced. The high-rate operation will produce a humus sludge mainly comprised of flocculated solids and detached fragments of film, whereas a low-rate filter sludge contains a large proportion of grazing fauna and animal fragments. Sludges containing animal fragments and grazing fauna will be more stable than sludges from high-rate systems where the grazing fauna is absent or reduced. The production of humus varies seasonally with mean production rates for low-rate filters between 0.20–0.25 kg humus per kg BOD removed. It varies from 0.1 kg $kg^{-1}$ in the summer to a maximum of 0.5 kg $kg^{-1}$ during the spring sloughing period. In high-rate systems, a greater volume of sludge is produced due to the shorter HRT, which results in less mineralization. The humus production does not vary seasonally to the same extent, having a mean production rate of 0.35 kg $kg^{-1}$, although this is dependent on the nature of the influent wastewater. Sludge production is much less compared with activated sludge, it being more stabilized and containing less water (Table 3.7), although as loading increases, the mode of purification in percolating filters approaches that of the activated sludge process, and the sludge alters

**Table 3.7.** Typical sludge volumes and characteristics from biological unit processes.

| Source | Volume (1 per head $d^{-1}$) | Dry solids (kg per head $d^{-1}$) | Moisture content (%) |
|---|---|---|---|
| Primary sedimentation | 1.1 | 0.05 | 95.5 |
| Low-rating percolating filtration | 0.23 | 0.014 | 93.9 |
| High-rating percolating filtration | 0.30 | 0.018 | 94.0 |
| Activated sludge (wasted) | 2.4 | 0.036 | 98.5 |

accordingly in quality and quantity. Separation takes place in secondary settlement tanks, traditionally called humus tanks. They are similar in principle to primary settlement tanks, although they are often of different designs. For example, deep square tanks of a small cross-sectional area with an inverted pyramid bottom with a maximum upflow rate of <1 m $h^{-1}$ are common (IWEM, 2000). Percolating filters usually achieve high levels of nitrification, and a long sludge retention time within the humus tank may give rise to anoxic conditions and problems of denitrification, which results in the carryover of sludge in the final effluent. High-rate sludges are more susceptible to denitrification than low-rate sludges, which are more stabilized and so exert less of an oxygen demand within the settlement tank. Temperature is also important with gas production being much heavier during the summer. Whereas the bulk of the solids settle easily, there is a fraction of fine solids that do not, and these are carried out of the humus tank in the final effluent. These fine solids are responsible for a significant portion of the residual BOD in the final effluent, and if a high-quality effluent is required, some form of tertiary treatment may be necessary. The residual BOD can be successfully removed by any tertiary treatment process, such as micro-straining, sand filtration, upflow clarification, or land treatment.

## 3.3 Process Modifications

Percolating filters originally were designed to treat domestic wastewater on a single pass through one filter bed containing mineral media of a nominal size between 25–50 mm. At a hydraulic loading of <0.4 $m^3m^{-3}d^{-1}$ and an organic loading of 0.12 kg BOD $m^{-3}d^{-1}$ a final effluent with a BOD and suspended solids concentration not exceeding 20 and 30 mg $L^{-1}$, respectively, can be obtained. Assuming a third of the organic load is removed during primary sedimentation, the per capita loading rate of a single-pass low-rate filter is equivalent to approximately 2.6 people per cubic meter of the filter medium. Over time, the general trend of increasing loads to treatment plants plus more stringent disposal standards has resulted in the necessity of expanding low-rate filters. However, because of cost and land availability restrictions, ways of uprating the efficiency of this original system have been examined. While there are a number of modifications in use that have effectively increased the per capita loading per unit volume of medium, the current trend is toward replacing or supplementing existing percolating filters. Common choices are RBCs and/or reed beds for small wastewater treatment plants (Chap. 5), submerged reactors for medium to large plants (Chap. 6), or activated sludge for large plants. Fixed film reactors remain a popular choice for industrial wastewaters and this has driven many innovative changes in the media (Sec. 3.2.3).

Filtration can be separated into single- and multi-stage processes. Wastewater treated by a single filter or by filters operated in parallel is known as single-stage filtration. Examples include low- and high-rate single filtration and recirculation. In multi-stage processes, the wastewater is treated progressively by filters operated in series and includes double filtration, alternating double filtration, and two-stage systems (Fig. 3.10).

### 3.3.1 *Single-Stage Systems*

Conventional single-pass low-rate filtration (Fig. 3.10) aims to produce a final effluent of a very high standard. With good operational conditions, almost complete purification including nitrification is possible. To achieve this, both the hydraulic and organic loadings must be kept low to obtain a low growth rate of film. Humus production is low, and the resulting sludge is highly mineralized and

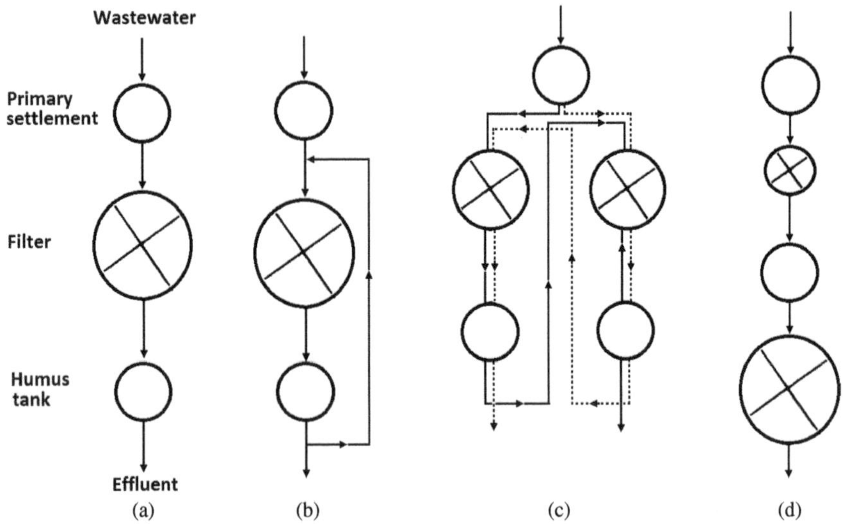


**Fig. 3.10.** Main systems of operation of percolating filters. (a) Single filtration, (b) recirculation, (c) alternating double filtration, and (d) two-stage filtration with high-rate primary filter.

easy to dewater. Using a medium of a nominal size of 37.5–50 mm, the maximum loading rate does not generally exceed 0.10–0.12 kg BOD $m^{-3}d^{-1}$ if a Royal Commission standard final effluent is required (i.e., 20 mg $L^{-1}$ BOD, 30 mg $L^{-1}$ suspended solids and full nitrification). In practice, the effluent BOD and suspended solids concentration fluctuate seasonally, with it being highest in spring. Such systems are employed at small sewage treatment plants, and where an appropriate hydraulic head is available they can be operated without a power supply. Where the flow is >1,000 $m^3d^{-1}$, modifications to conventional single-pass filters are usually required.

The earliest modification to the percolating filter process was recirculation, and it is now built into most single filtration systems (Fig. 3.10). Recirculation has generally been used to dilute strong wastewaters or wastewater containing a large proportion of industrial waste, particularly to reduce toxic components to below inhibitory concentrations, or occasionally to even out the diurnal fluctuations in flow to the filters, which ensures a uniform hydraulic loading. This modification results in an increase in the hydraulic loading to the filter, which has a number of advantages. The stronger flushing action prevents the surface accumulation of biofilm and reduces the biofilm accumulation within the bed, thus encouraging a thinner, faster-growing biofilm that is more efficient in removing organic matter. The greater hydraulic loading increases the area of the media utilized. The diluted influent also restricts biofilm accumulation at the surface of the filter because of a reduction in the available nutrients in the wastewater, whereas the recirculated fraction contains dissolved oxygen, nitrate and microorganisms that may be beneficial. Thus, by diluting the influent with recirculated treated effluent, higher organic loadings can then be applied to single-pass low-rate filters because the excessive biofilm accumulation is controlled. These factors allow the loading to be safely increased to between 0.15 and 0.20 kg BOD $m^{-3}d^{-1}$ with little effect on the final effluent quality (90–95% BOD removal), although nitrification will be slightly suppressed. When the nutrients in wastewater are soluble and readily oxidized, such as in food-processing wastes, the organic loading may be as high as 0.5–0.6 kg BOD $m^{-3}d^{-1}$. At higher hydraulic loadings, heterotrophic microorganisms utilize more of the filter depth because the food is more evenly distributed. This results in the zone of carbonaceous oxidation extending much deeper and restricts the area available for nitrification to the very base of the filter, occasionally excluding it altogether. The ratio of recirculated effluent to influent wastewater is the recirculation ratio. For domestic wastewater, the ratio is generally either 1:1 or 2:1. The recirculation ratio can rise to as high as 25:1 for very strong industrial wastewaters. The ideal loading rate after dilution should be between 0.1 and 0.15 kg BOD $m^{-3}d^{-1}$, although the actual effluent BOD concentration is normally used to determine the ratio, and therefore the organic loading is normally higher. The recirculation ratio can be calculated by estimating the BOD of the diluted influent fed to the filter ($S_o$):

$$S_o = (Q_i S_i + Q_r S_r)/(Q_i + Q_r),$$

where the flow rate ($Q$) and the BOD ($S$) are known for the influent (i) and the recycled or treated effluent (r). The recycle ratio $R_r$ is $Q_r$:$Q_i$, and the equation can be rearranged as:

$$S_o = (S_i + R_r S_r)/(1 + R_r).$$

Because the effluent BOD in low-rate systems is usually very low, $S_o$ can be estimated as:

$$S_o = S_i/(1 + R_r).$$

Recirculation can be considered as medium-rate filtration, whereas recirculation in high-rate systems has been found to have little effect on BOD removal efficiency. This is probably because the flushing action of the wastewater in plastic media filters is of less importance than in mineral beds.

Conventional single-pass filtration is also used for a partial treatment of wastewater (Fig. 3.10). Settled or raw wastewater that has been finely screened can be applied to filters containing mineral media with a large nominal size of >63 mm, random plastic medium, or more commonly modular plastic medium at high organic and hydraulic loadings. High-rate filtration can be used as a preliminary or roughing treatment of high strength industrial effluents for partial secondary treatment before discharge to a sewer or to surface water, especially estuaries and coastal waters, where the effluent discharge standards permit, and the first stage of secondary treatment followed by low-rate filtration or activated sludge (Fig. 3.11). The high voidage prevents ponding and heavy accumulations of film, so biofilm growth is rapid with constant and high removal of biofilm from the filter. The humus from such filters is poorly oxidized, making it less stable than conventional low-rate humus and more difficult to dewater (Bruce and Merkens, 1970). Almost any degree of partial treatment can be obtained (50–70% BOD reduction) depending on the loading rate. High-rate filters are operated at a hydraulic loading of >3 $m^3m^{-3}d^{-1}$ and at organic loadings of >0.6 kg BOD $m^3d^{-1}$. In general, loading 1 kg BOD $m^{-3}d^{-1}$ will give a BOD removal rate of between 80–90%, whereas this falls to 50% at loadings of between 3–6 kg BOD $m^{-3}d^{-1}$.

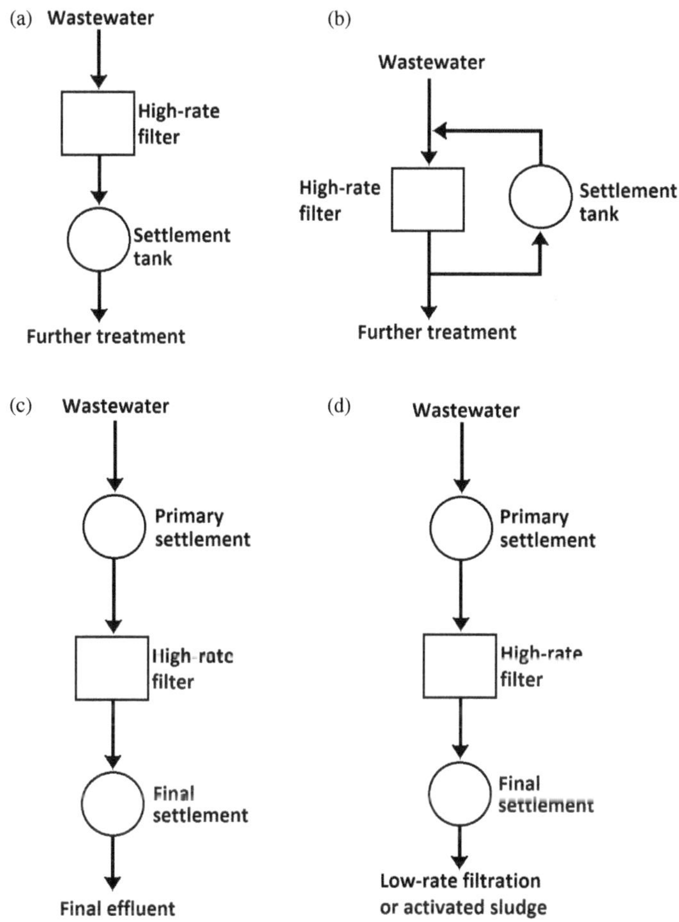


**Fig. 3.11.** Common applications of high-rate filtration: (a) Preliminary or roughing treatment of low solids industrial wastewater, (b) preliminary or roughing treatment of low solids industrial wastewater with partial recirculation, (c) partial secondary treatment of municipal or high solids industrial wastewater, and (d) first stage of multi-stage secondary treatment.

The specific removal rate depends on the nature of the wastewater, and generalizations are difficult. Performance data for a range of wastewaters is given elsewhere (Chipperfield, 1967; Jackson and Lines, 1972; Anderson, 1977). Coarse mineral media (63–150 mm) must be loaded at significantly lower organic loading rates than modular media. Modular and random plastic media have been successfully used to treat macerated and finely screened raw sewage. This eliminates the use of primary sedimentation, which results in a considerable saving in primary treatment and subsequent sludge handling (Hemming, 1979; Hoyland and Roland, 1984). The efficiency of high-rate filters is generally measured in terms of weight of BOD removed per unit volume of medium per day, rather than a percentage removal, as is the case with low-rate filters. Low-rate filters remove a high percentage of the influent organic matter, but the rate of removal is low in terms of a mass of nutrients removed per unit volume. Conversely, high-rate systems remove a smaller proportion of the influent organic matter but at a much higher rate. This has given rise to the general terms of "polishing" and "roughing" filters for low- and high-rate systems, respectively. The percentage removal of BOD decreases as the loading increases, whereas the mass of BOD removed increases with loading up to a critical limiting value above which the amount removed remains constant (Winkler, 1981). The advantages and disadvantages of low-rate and high-rate filtration are compared in Table 3.8.

### 3.3.2 *Multi-Stage Systems*

The most widely used multi-stage system is double filtration (DF). In this modification, two sets of similar beds are used in series. The media is essentially of the same grade (63–75 mm) and sedimentation is normally employed between each stage to minimize the risk of the humus solids clogging the interstices of the medium in the second filter. Intermediate sedimentation is not absolutely necessary, but if not used, it will reduce the performance slightly and increase the number of solids discharged from the second filter (Tomlinson and Hall, 1953). The major problem with using identical filters was that the primary filter often became ponded during winter, which severely limited the overall loading rate. This has been overcome with several modifications. Some degree of control of biofilm accumulation in the primary stage was achieved by using low-frequency dosing, although the system was vastly improved by using a high-rate medium in the first stage (two-stage filtration) and by alternating DF.

Two-stage filtration is a direct development of DF, although some textbooks fail to distinguish between the two. Here, different media

**Table 3.8.** Summary of the advantages and disadvantages of low- and high-rate filtration.

| **Advantages** | **Disadvantages** |
|---|---|
| *Low-rate (L-R) filtration* | |
| Well nitrified effluent | High land requirement |
| Natural ventilation adequate | Effluent sometimes turbid |
| No aerosols or foams | Fly nuisance may occur |
| Low sludge yield | No biological removal of N or P |
| No sludge recycling | Seasonal sloughing of film |
| Low operating costs | High capital cost |
| Resistant to shock loads | Final effluent BOD >10 mg $L^{-1}$ |
| No or low power requirement | Final effluent suspended solids >10 mg $L^{-1}$ |
| *High-rate filtration* | |
| Lower capital cost than L-R filtration | High BOD and suspended solids in effluent |
| High BOD removal per unit volume | No nitrification |
| Less prone to fly nuisance | Main source of odor at treatment plant |
| Lower area required than L-R filters | Sludge more difficult to dewater Higher sludge yield than L-R filters |
| | Often visibly obtrusive |
| | Power required |

are used in each filter with the first stage operated as a roughing filter, and the second stage for polishing the partially treated wastewater. The first stage is operated as a high-rate filter at high loading rates using either a coarse mineral medium (75–130 mm) or plastic media with a high voidage. This first stage reduces the organic loading by 70%, thus allowing the second stage, after intermediate sedimentation, to be operated as a conventional low-rate filter. At loading rates of between 1.6–2.3 kg BOD $m^{-3}d^{-1}$ for the first filter, and 0.04–0.12 kg BOD $m^{-3}d^{-1}$ for the second stage, a 20:30 effluent with a high degree of nitrification is possible (Forster, 1977). The system can be further enhanced by using recirculation. Where very strong wastewaters are being treated, it may be necessary to have several roughing filters in series to reduce the organic loading sufficiently to allow complete treatment by the final low-rate stage. This multiple unit system is known as multi-stage filtration (Fig. 3.11).

A further development of the DF system is alternating double filtration (ADF). Identical filters are used, each containing the same medium, with the biofilm accumulation in the first stage controlled by periodically reversing the sequence the filters are used so that each filter is subject to successive periods of feeding and starvation (Fig. 3.10). The filters are filled with a coarser medium (63–75 mm) than normally used for single filtration, which allows a relatively high loading to the filter. The biofilm is controlled to a certain extent by the high flow, distributing the growth of biofilm more evenly through the filter and providing some restriction of biofilm accumulation by washing out the film. The rate of biofilm growth at the high organic loadings used is very rapid in the first filter of the sequence. Loading in this order continues for 1–2 weeks until the filter has accumulated a heavy growth of biofilm and is close to becoming clogged. At this point, the flow is reversed so that the second filter is being fed with dilute partially purified wastewater. This puts the biofilm in the second filter under substrate-limited conditions and forces the microorganisms to enter an endogenous growth phase, which reduces the overall biofilm biomass, but at the same time retains a healthy and active biofilm. Humus is removed after every stage, and the ADF process allows the organic loading to be increased by at least a factor of two. For domestic wastewater, a 20:30 effluent can be produced at a loading of 0.24 kg BOD $m^{-3}d^{-1}$ (1.5 $m^3m^{-3}d^{-1}$) (Calley *et al.*, 1977), although loadings of 0.30–0.47 kg BOD $m^{-3}d^{-1}$ are normal (Forster, 1977). The starvation effect on the second filter is important to ensure biofilm reduction, and with stronger wastes, such as meat-processing wastewaters, recirculation may also be necessary to achieve wastewater diluted enough for the second stage. A major disadvantage with the ADF process is that nitrification is suppressed. Nitrifying bacteria are normally found at the base of conventional filters where all the carbonaceous material has been utilized. The nitrifying bacteria are slower growing than heterotrophs and are unable to compete for available space. Thus, by alternating the filters in this way, nitrifying bacteria fail to become well established, which results in poor nitrification (Bruce *et al.*, 1975).

**Table 3.9.** Summary of the percolating filter operating variables: (I) is the first stage, and (II) is the second stage of treatment.

| Key operating variables | Low-rate | Recirculation | Double filtration |
|---|---|---|---|
| Medium type | Blast furnace slag | Blast furnace slag | Blast furnace slag |
| Medium grade (mm) | 50 | 28–40 | (I) 100–150, (II) 28–50 |
| Medium depth (m) | 1.8 | 1.8 | (I) 1.8–7.0, (II) 1.8 |
| Organic loading rate (kg BOD $m^3$ $d^{-1}$) | 0.07–0.15 | 0.10–0.20 | 0.10–0.20 |
| Maximum influent BOD (mg $L^{-1}$) | <350 | <2,000 | <2,000 |
| Recirculation ratio | 0 | 1:1 to 5:1 | 0 |

Modern percolating filter plants are designed with enough flexibility to allow the filters to be operated at single, double, or even ADF filtration, depending on the strength of the incoming wastewater and the degree of treatment required. Basic operating parameters are summarized in Table 3.9.

### 3.3.3 *Modeling*

There are many simple models available to help in the design of both low- and high-rate filters (Pike, 1978; Hoyland and Horwood, 1980; Martin and Anderson, 2007). More dynamic mathematical models are based on the Michaelis–Menton equation to measure the utilization of substrate within the film, the Monod equation to estimate biofilm growth and detachment, and Fick's law for substrate diffusion through the film.

Empirical design equations or models are based on the performance data collected from either full-scale or pilot plant systems. Initial models were for mineral media trickling filters, although specific models have been developed for all FFR designs, including trickling filters using plastic media. The National Research Council model below, which is still widely used, is based on the observation of 34 domestic single-stage trickling filters (NRC, 1946).

$$E_1 = \frac{100}{1 + 0.4432\sqrt{\frac{W_1}{VF}}},$$

where ***E*** is the BOD removal (%), ***W*** is the organic loading rate (Kg BOD $d^{-1}$), ***V*** is the volume of media ($m^3$), and ***F*** is the recirculation factor.

The recirculation factor F is calculated as:

$$F = \frac{1 + R}{(1 + R/10)^2},$$

where ***R*** is $Q_r/Q$ (i.e., the recirculation flow/wastewater flow).

## Example:

Calculate the dimensions of a single-stage mineral medium trickling filter required to treat the wastewater from a small town with a population equivalent of 2,500 with a per capita wastewater production of 180 liters per day. The strength of the influent wastewater is 320 mg $L^{-1}$ with a final effluent quality of 20 mg $L^{-1}$ required with a recirculation ratio of 1:1.

First, calculate the required efficiency of BOD removal required ($E_1$):

$$E_1 = (320 - 20/320) \times 100 = 83\%.$$

Next, calculate the BOD loading ($W_i$):

$$W_i = 2500 \times 180 \times 320/10^6 = 144 \text{ kg d}^{-1}.$$

The recirculation ratio (F) of 1:1 for a single-pass filter is:

$$F = \frac{1 + R}{(1 + R/10)^2} = \frac{1 + 1}{(1 + 1/10)^2} = -1.65.$$

Finally, apply the NRC model:

$$E_1 = \frac{100}{1 + 0.4432\sqrt{\frac{W_1}{VF}}},$$

$$83 = \frac{100}{1 + 0.4432\sqrt{\frac{144}{V1.65}}},$$

$$V = 408.94 \text{ m}^3.$$

Therefore, to treat the wastewater, a mineral filter 17 m in diameter and 1.8 m in depth is required to give a total medium volume equivalent to 408.94 m³.

If the recirculation ratio is increased to 2:1, then:

$$F = \frac{1+2}{(1+2/10)^2} - 2.083,$$

$$83 = \frac{100}{1 + 0.4432\sqrt{\frac{144}{V2.083}}},$$

$$V = 323.69 \text{ m}^3.$$

Therefore, the volume of the filter required is reduced, and the reactor is smaller with a reduction in capital costs.

Where there are two stages, the second stage is modeled as:

$$E_2 = \frac{100}{1 + \frac{0.4432}{1 - E_1}\sqrt{\frac{W_2}{VF}}}.$$

With the final effluent after two stages, ($\boldsymbol{E_2}$) is expressed as a percentage.

The majority of biological treatment systems operate in the mesophilic temperature range (20–40°C). However, like activated sludge systems, most FFRs operate at the temperature of the wastewater (i.e., 12–25°C), although the air temperature and the rate of ventilation in trickling filters can have a profound effect on heat loss (Sec. 3.4.2). The higher temperatures result in increased biological activity that, in turn, increases the rate of substrate removal. The increased metabolism at higher temperatures can also lead to problems of oxygen limitation. Generally, activated sludge systems perform better than trickling filters below 5–10°C, although heterotrophic growth continues at these temperatures. However, the practice in colder climates of covering filters and controlling the rate of ventilation, thus reducing heat loss within the filter bed, has largely overcome this problem. Van't Hoff's rule states that the rate of biological activity doubles with every 10°C rise in temperature within the range of 5–35°C. The variation in reaction rate is represented by the modified Arrhenius expression:

$$k_1 = k_{20}\phi^{T-20},$$

where **T** is the temperature (°C), ***k*** is the reaction rate constant at temperature T ($d^{-1}$), $\boldsymbol{k_{20}}$ is the reaction rate constant at 20°C ($d^{-1}$), and $\boldsymbol{\phi}$ is the temperature coefficient. There is a rapid decrease in the growth rate as the temperature increases above 35°C, which falls to zero as the temperature approaches 45°C. Therefore, the efficiency of trickling filters is temperature dependent with performance falling as temperature declines. So, filter efficiency can be adjusted for temperature, using the equation:

$$E_t = E_{20}\,\phi^{T-20},$$

where $\boldsymbol{E_t}$ is the filter efficiency at temperature **T**, $\boldsymbol{E_{20}}$ is the filter efficiency at 20°C, and $\boldsymbol{\phi}$ is the temperature coefficient, which is 1.035.

Another widely used empirical model is used in various UK manuals of practice:

$$\frac{L_i}{L_o} = \frac{1}{\left[1 + k\theta^{T-15}\left(\frac{As^m}{Q^n}\right)\right]},$$

where $L_i$ is the influent BOD concentration (mg $L^{-1}$), $L_o$ is the settled effluent BOD concentration (mg $L^{-1}$), and $k$ is the rate constant (ranging from 0.02 for blast furnace slag to 0.4 for random plastic media) (Sec. 3.2.3), $As$ is the specific surface area of the medium ($m^2m^{-3}$), $m$ is the associated constant (1.41 for mineral and 0.73 for plastic media), $Q$ is the volumetric loading rate ($m^3m^{-2}d^{-1}$), $n$ is the associated constant (1.25 for mineral and 1.40 for plastic media), $\phi$ is the temperature coefficient (1.111), and $T$ is the wastewater temperature (°C).

A number of empirical models were developed in the UK during the 1970s and early 1980s. Lamb and Owen (1970) produced a simple empirical model, although no account of filter depth or the effects of recirculation were included.

$$\frac{Ce}{Ci - Ce} = \frac{7.2Q}{VAs}e^{-0.09(T-15)},$$

where $Ci$ is the influent BOD concentration (mg $L^{-1}$), $Ce$ is the settled final effluent BOD concentration (mg $L^{-1}$), $Q$ is the influent flow rate ($m^3d^{-1}$), $V$ is the volume of media, $As$ is the specific surface area of the medium ($m^2m^{-3}$), and $T$ is the wastewater temperature.

The Water Research Centre ran 46 different evaluations, comprising 35 different combinations of media types at a range of hydraulic loadings, over 8 years in large single-stage trickling filters (Pike, 1978). They produced a series of simple empirical-based equations that allow engineers to easily study the factors most likely to influence performance (i.e., the specific surface area of the media, hydraulic loading and temperature) based on the model:

$$C_e/C_i = 1/(1 + K\,\phi^{(T-15)}\,S^m/Q^n),$$

where $\mathbf{C_e/C_i}$ is the fraction of BOD remaining after settlement, **T** is the temperature of the wastewater (°C), **S** is the specific surface area of the media ($m^2m^{-3}$), **Q** is the hydraulic loading rate ($m^3m^{-3}d^{-1}$), **K** is the first-order rate coefficient, $\boldsymbol{\phi}$ is a temperature coefficient, **m** is a coefficient of the wetting of the medium, and **n** is a coefficient relating to the physical characteristics of the medium. The resultant equations apply to a wide range of hydraulic loadings (0.3–16.0 $m^3m^{-3}d^{-1}$), temperatures (9–21°C), and media-specific surface areas (40–211 $m^2m^{-3}$). So, for the removal of BOD from single-stage mineral or randomly packed plastic media:

$$C_e/C_i = 1 / (1 + 0.0204 \times 1.111^{(T-15)} \times S^{1.407}/Q^{1.249}).$$

For modular plastic media:

$$C_e/C_i = 1 / (1 + 0.4 \times 1.089^{(T-15)} \times S^{0.7324}/Q^{1.396}).$$

From this study the removal efficiency of anionic surfactants expressed as the fraction of the applied surfactant remaining $\mathbf{(D_e/D_i)}$ could be estimated using the equation below, which is suitable for both conventional and high-rate trickling filters:

$$D_e/D_i = 1 / (1 + 0.0608 \times 1.046^{(T-15)} \times S^{1.097}/Q^{1.735}).$$

Ammonia removal $\mathbf{(N_i–N_e)/N_i}$ was estimated by using the linear equations:

Where $Q/S > 0.0165$:

$$(N_i\text{-}N_e)/N_i = 0.06125 - 0.3551\ (Q/S) \times 0.9452^{(T-15)}.$$

If $Q/S < 0.0165$:

$$(N_i\text{-}N_e)/N_i = 1.0404 - 53.871\ (Q/S) \times 0.9452^{(T-15)}.$$

This study formed part of the Water Research Centre's major model for the evaluation of sewage treatment plants called *STOM*

(Sewage Treatment Optimization Model), which covered nearly all trickling filter configurations and types except for recirculation systems (Water Research Centre, 1981).

Velz (1948) developed the kinetic approach that is more commonly used today. This model applies to single-pass filters, assuming the first-order removal of COD, thus making COD removal a function of the HRT. This modified equation is primarily used for plastic media:

$$\frac{S}{S_o} = e^{-}\left[\frac{k_{20} A_s D\theta^{T-2\theta}}{Q^n}\right].$$

This model was subsequently modified by Eckenfelder (1980) by including the influent substrate concentration ($S_o$):

$$\frac{S}{S_o} = e^{-}\left[\frac{k_{20} A_s D\theta^{T-2\theta}}{Q^n S_o}\right].$$

So the COD removal is a function of the organic loading rate:

$$\frac{S}{S_o} = e^{-}\left[\frac{k_{20} A_s \theta^{T-2\theta}}{L}\right].$$

Kincannon and Stover (1982) proposed an alternative, the total organic loading model, which is based on surface organic loading rates. This model is similar to the Strover–Kincannon model for RBCs:

$$\frac{S_o - S}{S_o} = \frac{U_{max}}{L' + K_b},$$

where $\boldsymbol{A_s}$ is the specific surface area of the medium ($m^2m^{-3}$), $\boldsymbol{D}$ is the depth of the filter (m), $\boldsymbol{k_{20}}$ is the treatability constant, $\boldsymbol{K_b}$ is a biokinetic parameter based on media area ($g\ m^{-2}d^{-1}$), $\boldsymbol{L}$ is the organic loading rate (kg sCOD

$m^{-3}d^{-1}$), ***L'*** is the surface organic loading rate (g sCOD $m^{-2}d^{-1}$), ***n*** is the hydraulic flow exponent (0.5), ***Q*** is the hydraulic loading ($m^3m^{-2}d^{-1}$), ***S*** is the effluent sCOD concentration (mg $L^{-1}$), sCOD is the soluble Chemical Oxygen Demand (mg $L^{-1}$), $S_o$ is the influent sCOD concentration (mg $L^{-1}$), and $U_{max}$ is a biokinetic parameter (g $m^{-2}d^{-1}$).

Raj and Murthy (1999) compared these kinetic models using pilot-scale trickling filters with a crossflow PVC medium to treat synthetic dairy wastewater (COD: 1,240 mg $L^{-1}$, BOD: 950 mg $L^{-1}$, total nitrogen: 66 mg N $L^{-1}$, and total phosphorous: 12 mg P $L^{-1}$). They operated their filters over a range of different organic loading (0.54–4.45 kg sCOD $m^{-3}d^{-1}$) and hydraulic loading (4–17 $m^3m^{-2}d^{-1}$) rates, and influent sCOD concentrations (195–471 mg $L^{-1}$). The results showed that organic removal rates increased with increasing organic loading up to a threshold of 2.36 kg sCOD $m^3d^{-1}$, after which the rate of removal slowly declined until a constant (Fig. 3.12). Comparing their results to those predicted by the models, it was the model of Kincannon and Strover that gave the best agreement by far.

Schulze, Germain, Galler, Logan, Gotaas and many more researchers have all produced models for trickling filters (Logan *et al.*, 1987; Reyes-Lara and Reyes-Mazzoco, 2009; WEF, 2011). However, the most commonly used model for plastic media is the Schulze formula, which is based on the assumption that the time the wastewater is in contact with the biofilm is proportional to the depth but inversely proportional to the HRT. So:

$$t = \frac{CD}{q^n},$$

where ***t*** is the time that the wastewater is in contact with the biofilm (d), ***D*** is the depth of the medium (m), ***q*** is the surface hydraulic loading rate ($m^3m^{-2}d^{-1}$), while ***C*** and ***n*** are constants related to the specific surface and configuration of the media. The surface hydraulic loading rate (***q***) is calculated as:

$$q = \frac{Q(1+R)}{A},$$

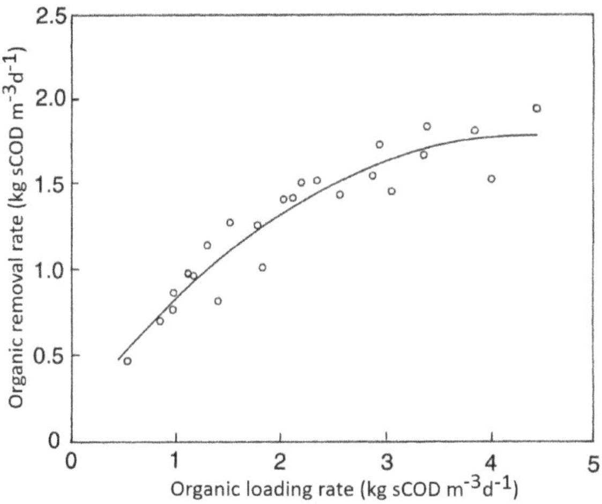


**Fig. 3.12.** The effect of increasing the organic loading rate on removal efficiency (Raj and Murthy, 1999). Reproduced with permission of Springer Nature.

where ***Q*** is the influent flow rate ($m^3d^{-1}$), ***R*** is the recirculation ratio, and ***A*** is the filter cross-sectional area ($m^2$). The Schulze formula is expressed as:

$$\frac{S}{S_o} = e^{\left(-kD/q^n\right)},$$

where ***S*** is the BOD concentration of settled effluent (mg $L^{-1}$), $S_o$ is the influent BOD concentration (mg $L^{-1}$), ***k*** is the wastewater treatability coefficient (0.69 $d^{-1}$ @ 20°C), and ***n*** is the hydraulic constant of packing used (assumed to be 0.67).

## 3.4 Factors Affecting Performance

A number of factors that affect the performance of percolating filters have already been discussed in some detail, especially the effect of hydraulic and organic loadings. It is important, of course, that the wastewater must be amenable to treatment by filtration and have a satisfactory C:N:P ratio for aerobic oxidation (100: ≥ 5: ≥ 1) (Gray,

2004). Other important factors include the media, temperature, retention time, depth, film, oxygen, and frequency of dosing, all of which are dealt with below.

### 3.4.1 *Media*

Performance is generally related to the specific surface area of the medium, provided biofilm accumulation does not result in blocking of the interstices. Therefore, smaller grades of media will produce a better quality final effluent than a larger grade of the same medium, when loaded at the same rate (Table 3.10) (Truesdale *et al.*, 1962). The potential for biofilm accumulation is an important selection criterion for media, which is dependent on the size of the interstices, the specific surface area, surface texture of the medium, and the organic loading. Smoother media are slightly less effective than rough-surfaced media, although the latter is more susceptible to excessive accumulation of solids. Performance varies seasonally with the larger grades of media producing a better effluent in the winter, whereas the smaller grades perform best in the summer (Hawkes and Jenkins,

**Table 3.10.** Mean results of chemical analysis over 12 months of influent-settled sewage and final (settled) effluent from percolating filters containing different types and sizes of media. The mean rate of the application of sewage was 0.59 $m^3m^{-3}d^{-1}$. The specific surface area ($m^2m^{-3}$) of the bulk medium, as determined by the paint dipping technique are given in brackets (Truesdale *et al.*, 1962). Reproduced with permission of the Chartered Institution of Water and Environmental Management.

| | Influent | Settled effluent from filter containing | | | | | | | |
|---|---|---|---|---|---|---|---|---|---|
| | | 25 mm nominal size | | | | 65 mm nominal size | | | |
| | | Clinker (202) | Slag (196) | Rock (142) | Rounded (gravel) (146) | Clinker (123) | Slag (108) | Rock (91) | Rounded (gravel) (65) |
| BOD (mg $L^{-1}$) | 308 | 12 | 10 | 18 | 21 | 18 | 21 | 26 | 28 |
| Ammonia (mg N $L^{-1}$) | 61 | 4.0 | 1.8 | 13.1 | 13.5 | 13.3 | 18.6 | 32.5 | 40.6 |
| Oxidized nitrogen (mg N $L^{-1}$) | — | 52.7 | 50.7 | 43.3 | 44.0 | 40.1 | 34.2 | 24.7 | 18.3 |

1955). Therefore, the selection of media must be a compromise between having sufficient specific surface area and large enough interstices to prevent clogging in the winter. The effects of the media type on performance have already been considered in an earlier section (Sec. 3.2.3).

### 3.4.2 *Temperature*

Temperature is an important operational parameter in all biological wastewater treatment systems. In percolating filters, the metabolic rate of microorganisms increases with temperature, with the rate of treatment doubling for every 10°C increase within the range of 5–30°C. However, the diffusivity of nutrients and oxygen increases with temperature and the solubility of oxygen decreases. As the BOD removal efficiency ($E$) falls with declining temperature (T) the relationship can be expressed as:

$$E_T = E_{20}.A_T^{(T-20)},$$

where $\boldsymbol{E}_T$ and $\boldsymbol{E}_{20}$ are the BOD removal efficiencies at T°C and 20°C, respectively, and $A_T$ is a temperature coefficient with a value between 1.035–1.047 (Roberts, 1973). The temperature affects biofilm accumulation if the rate of oxidation and the activity of grazers are reduced. This is severely restricted at temperatures below 10°C (Hawkes 1957; Solbe *et al.*, 1974; Shephard and Hawkes 1976). Nitrification is also temperature-dependent with little activity below 10°C (Bruce *et al.*, 1967) (Fig. 3.13).

The normal range of domestic wastewater is between 10–20°C, although Gray (1980) recorded a range of 6.5–18.0°C for settled influent wastewater loaded onto his pilot filters (Table 3.11). Thus the design loading for a percolating filter is restricted by the lowest winter temperature. The temperature of the influent wastewater and the biofilm varies seasonally (Fig. 3.14). Gray (1980) examined the effect of the temperature on percolating filter performance in his pilot filters, which were built above ground. The concrete walls, although thick enough to offer some insulation, allowed heat transfer between

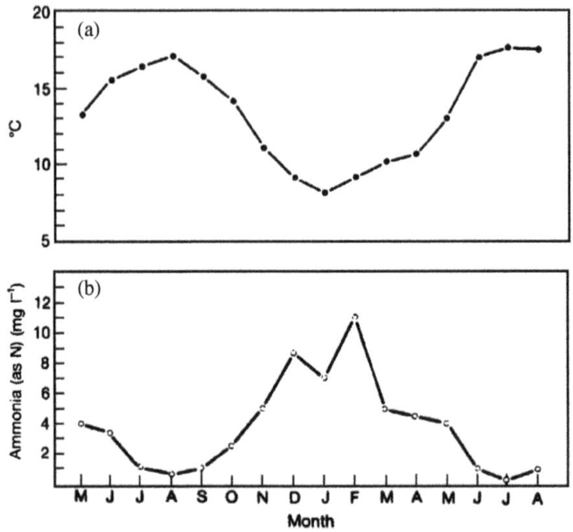


**Fig. 3.13.** Seasonal variation in the temperature of sewage applied to a percolating filter and the concentration of ammonia remaining in the effluent. The average concentration of ammonia over the whole period was 5.1 mg N $L^{-1}$. (a) Monthly average temperature of the applied sewage, and (b) monthly average concentration of ammonia in the filter effluent (Bruce *et al.*, 1967). Reproduced with permission of the Chartered Institution of Water and Environmental Management.

the filter medium and the air outside. Changes in the air temperature, wind velocity affecting heat loss from the walls, and the absorption of heat by solar radiation all caused temperature changes to occur in the medium immediately adjacent to the walls within the filters. The thermocouples inside the pilot filters closely recorded these temperatures (Gray and Learner, 1983). There was a clear diurnal variation in the temperature in the medium immediately adjacent to the wall — it got warmer during the day and cooler at night. The temperature of the central core of the filters remained far more constant. Gray compared 50 mm of blast-furnace slag with a random plastic filter medium of the same nominal size (Flocor RC). He found that, in fact, the core temperature in the slag medium, as measured by the central thermo-couple, remained extremely constant with small changes in temperature occurring over long periods, i.e., over six hours, whereas the temperature in the plastic medium changed more rapidly, often by 1°C within 30 minutes. During the winter, the temperature gradually increased with depth due to the heat produced from microbial activity, although the medium adjacent to the walls of the filters remained cooler. In the summer, the influent retained more of its original heat and the greater microbial activity meant that the temperature of the influent sewage again increased with depth but to a greater degree than before. The metabolic rate increases exponentially with temperature. Therefore, greater temperatures are to be expected during the warmer months. The temperature of the slag filter was closely related at both low- and high-rate operations to the temperature of the influent. The air temperature affected the temperature of the influent wastewater (Fig. 3.14). During extremely cold conditions, the influent was also cooled considerably in the fall from the distributor to the surface of the medium.

Although the temperature of the surface layer of the medium was influent, the core temperature of the filter remained constant due to heat production from biological oxidation. The slag filter comprised of 48.5% solid material compared to only 8.7% in the plastic filter, with the mineral media able to retain heat more effectively due to a higher bulk density. Therefore, with more

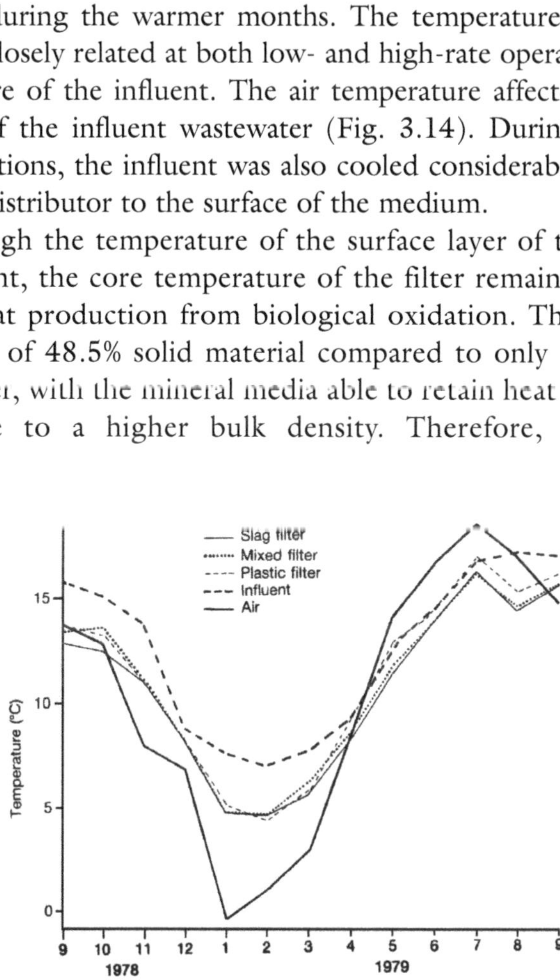


**Fig. 3.14.** Seasonal changes in influent, final effluents and air temperatures during high-rate loading of pilot percolating filters that are 1.8 m deep and loaded at 3.37 $m^3m^{-3}d^{-1}$.

heat-retaining material, less voidage, and much lower potential ventilation than the plastic filter, the slag filter was a far more effective buffer against changes in the air temperature; the central core of the filter being maintained at a constant temperature by the heat produced by biological oxidation. The temperature of the plastic medium filter was influenced far more by the air temperature than that of the influent because of the greater voidage, giving rise to excessive, and therefore more heat exchange. The plastic filter was seen to respond quickly to changes in air temperature. This close association resulted in variations in the diurnal temperature of the biofilm that were not reflected in the seasonal data (Table 3.11). Bayley and Downing (1963) suggested that with high voidage synthetic media, the air temperature and rate of the flow of air influenced the temperature in the voids, but that the temperature of the influent was still the main factor in maintaining the temperature, and to a lesser extent, the rate of reaction within the microbial film. The temperature of the influent prevents extremes of temperature within the plastic filter and this can be seen clearly at the higher loading. In tall towers containing modular plastic media, it is necessary to control the rate of ventilation through the filter — by adjusting the air vents at the base — to maintain the temperature, especially on cold windy days to prevent excessive cooling.

Warm wastewaters are advantageous in reducing seasonal fluctuations in temperature, which prevents low temperatures in the winter. In colder climates, filters are covered to conserve heat and to prevent freezing. Gebert and Wilderer (2000) compared two filters at the Ingolstat Wastewater Treatment Works in Germany to study the effects of warming the medium to increase the treatment capacity of filters during cold weather conditions. By embedding heat exchange pipes into the medium of one filter and comparing it to an identical unheated filter, they found that the heated filter did not consistently produce a better final effluent. They concluded that the decrease in oxygen solubility in the wastewater, coupled with mass transfer limitations caused by an increase in biofilm thickness in the heated filter, as the main reason for the lack of improvement in overall efficiency compared to the unheated filter.

**Table 3.11.** Summary of the mean performance of three pilot filters over a range of hydraulic loadings containing 50 mm blast furnace slag (Slag), 50 mm random plastic medium (Flocor RC™) (Plastic) and 50 mm blast furnace slag with a later 50 mm random plastic medium on the surface (Mixed).

| Loading (duration) | 1.68 $m^3m^{-3}d^{-1}$ (13 months) | | | 3.37 $m^3m^{-3}d^{-1}$ (13 months) | | | 5.72 $m^3m^{-3}d^{-1}$ (3 months*) | | |
|---|---|---|---|---|---|---|---|---|---|
| Filter | Slag | Mixed | Plastic | Slag | Mixed | Plastic | Slag | Mixed | Plastic |
| Organic load (kg BOD $m^{-3}d^{-1}$) | 0.280 | 0.280 | 0.280 | 0.628 | 0.628 | 0.628 | 0.854 | 0.854 | 0.854 |
| Effluent BOD (mg $L^{-1}$) | 20.20 | 20.53 | 22.54 | 33.09 | 25.03 | 27.12 | 33.70 | 26.25 | 32.00 |
| Percentage removal BOD | 87.58 | 87.73 | 86.26 | 82.36 | 87.18 | 85.50 | 68.00 | 74.50 | 68.60 |
| Suspended solids load (kg $m^{-3}d^{-1}$) | 0.201 | 0.201 | 0.201 | 0.417 | 0.417 | 0.417 | 0.743 | 0.743 | 0.743 |
| Effluent suspended solids (mg $L^{-1}$) | 24.73 | 24.72 | 27.53 | 31.08 | 26.32 | 28.79 | 71.00 | 39.50 | 27.00 |
| Percentage removal suspended solids | 77.82 | 77.67 | 75.66 | 72.20 | 77.00 | 78.30 | 41.60 | 60.10 | 66.40 |
| Ammonia load (kg $m^{-3}d^{-1}$) | 0.054 | 0.054 | 0.054 | 0.075 | 0.075 | 0.075 | 0.114 | 0.114 | 0.114 |
| Effluent ammonia (mg $L^{-1}$) | 14.71 | 17.80 | 20.73 | 17.98 | 15.75 | 19.09 | 16.90 | 16.60 | 16.70 |
| Percentage removal $NH_3$ | 52.81 | 44.97 | 34.86 | 21.08 | 32.42 | 17.75 | 7.75 | 8.45 | 8.2 |
| Total oxidized nitrogen (mg $L^{-1}$) | 14.32 | 12.09 | 9.03 | 6.19 | 6.95 | 4.30 | 1.05 | 1.15 | 0.75 |
| Effluent temperature (°C) | 8.97 | 8.78 | 8.94 | 10.86 | 10.64 | 11.09 | — | — | — |

*These results were collected over a shorter period during maturation of the filters, and therefore are not directly comparable to the results collected during the other loadings.

### 3.4.3 *Retention Time*

The longer the intimate contact is maintained between wastewater and the active biofilm supported on the medium of percolating filters, the better the final effluent quality will be. Clearly, the duration of liquid retention within the filter is potentially an extremely useful and important parameter. However, prediction of the performance of filters using the retention time has proved to be largely unsuccessful. Although retention time is considered a major factor in filter efficiency (Tariq, 1975), the exact meaning of such data has remained unclear, resulting in its infrequent use in the assessment of filter efficiency.

The importance of retention time, also referred to as residence or contact time, in the assessment of the efficiency of filters has been stressed from the earliest of times (Royal Commission on Sewage Disposal, 1908). The theory that increased retention allows more time for the wastewater to be in intimate contact with the film, therefore allowing greater adsorption of particulate matter and maximum exchange of nutrients, has been widely studied (Craft and Ingols, 1973; Cook and Katzberger, 1977). Many of these workers found that the retention time was associated with, although not directly related to, performance. Eden *et al.* (1964), when examining the measurement and significance of retention time in filters, found that there were three controlling factors: the hydraulic flow, biofilm accumulation, and the size and shape of the filter medium. As the physical characteristics of the medium remain constant, any variation in the retention time of a particular filter will be due to diurnal or daily changes in the hydraulic flow or seasonal changes in the biofilm accumulation. Increases in hydraulic flow normally result in a decrease in retention time, except in certain random plastic filter medium where an increase in hydraulic loading results in better redistribution of the wastewater within the bed, thus maintaining the retention time (Bruce *et al.*, 1975; Porter and Smith, 1979). Escritt (1965) found a strong inverse relationship between the hydraulic loading and retention time in a conventional full-scale filter of 1.8 m depth. He was able to calculate the retention time ($t_r$) in minutes as:

$$t_r = 40\ \mathrm{In}(9.4/L_v),$$

where $L_v$ is the hydraulic retention time ($m^3m^{-3}d^{-1}$). Generally, the fluctuations in hydraulic flow to filters are small and will only have slight effects on the retention time. The effects of biofilm accumulation on retention time are less clear. As no direct relationship has been established, it would appear that the biofilm modifies the flow pattern of the wastewater at particular accumulations. This was demonstrated by Gray (1983b), who showed that biofilm and humus had a pronounced effect on retention time, but that retention time was not directly related to removal efficiency. Large variations in the retention time had little effect on the organic removal efficiency of low-rate filters, whereas small variations in retention time had large effects on the removal efficiency of high-rate filters.

The results of retention tests are commonly expressed as the time (in minutes) required for the recovery of 16% ($t_{16}$) and 50% ($t_{50}$) of the added tracer, taken from the plot of percentage tracer recovered against time on logarithmic paper, on which a log-normal distribution is a straight line. The 16th percentile and 50th percentile (or median) values are chosen as ($t_{50}/t_{16}$) = $\sigma$, the standard deviation of the log-normal distribution (Eden *et al.* 1964). Gray used $t_{50}/t_{16}$ to predict the flow pattern of wastewater within the filter. He found that when channels are formed in filters because of excessive biofilm accumulation, some of the tracer passed through the filter extremely quickly, causing the rate of discharge to decrease after a short period, and producing a ratio ($t_{50}/t_{16}$) of >3.125. Under this biofilm condition, and during periods of normal biofilm accumulation, the median retention time (MRT) increased as the biofilm accumulation increased, with a ratio at 3.125 or less. It was during this period that the MRT was directly associated with the performance. When the MRT increased rapidly to a very high value, and the ratio became more than 3.125, this indicated that the influent was being retained in reservoirs formed by excessive biofilm conditions. In this instance, although the MRT is longer, it is not associated with increased performance. Because the influent is stored in reservoirs only a very small proportion of the liquid is in intimate contact with the film, thus nutrient transfer does not take place so efficiently. Also, wastewater stored in this way will be low in dissolved oxygen and effectively eliminates

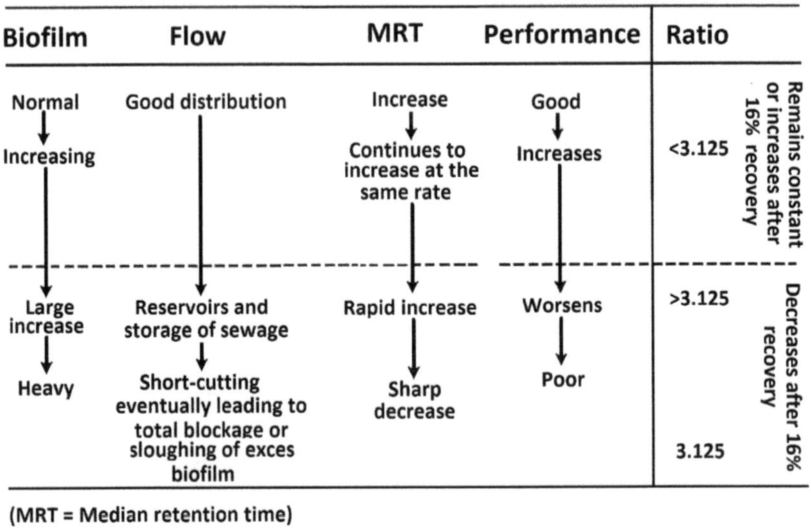


**Fig. 3.15.** Summary of observed flow characteristics in relation to the ratio of the 16th to the 50th percentiles $\left(\frac{t_{50}}{t_{16}}\right)$ of the retention time in pilot percolating filters.

oxygen transfer into the film, severely reducing aerobic metabolic activity. An increase in biofilm accumulation either leads to the production of channels, or if the excess biofilm is not sloughed or removed by the grazing fauna, the filter becomes completely blocked. These processes are summarized in Fig. 3.15.

The size, shape and surface area of the media are also important, with the MRT increasing with the specific surface area of the media (Fig. 3.16). However, as the medium within a filter remains constant, it can be ignored in terms of operation. Work on the periodicity of dosing has shown that retention time can be increased by large intermittent doses of influent (Shephard, 1967) due to the storage of liquid in horizontal chambers within the filters. These reservoirs are found in filters at times of high accumulations of biofilm where the tracer enters quickly but is slowly discharged as more influent wastewater replaces and dilutes it. This explains why the large increase in MRT recorded before the rapid decline in MRT is not directly associated with performance. In contrast, a number of workers consider

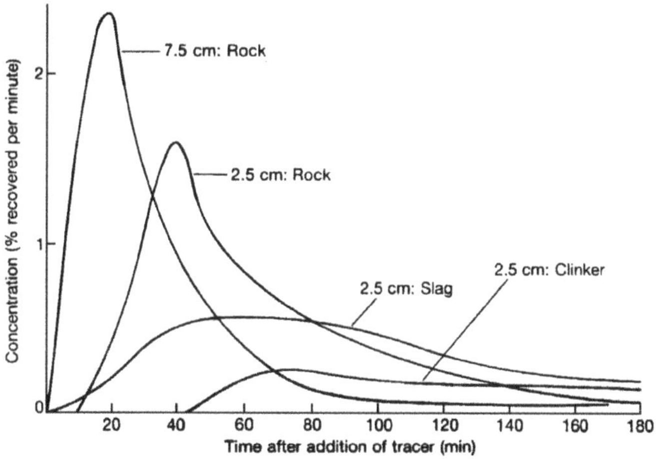


**Fig. 3.16.** Comparison of the retention characteristics of various clean media (i.e., without biofilm) in pilot filters that are 1.8 m deep and loaded at 0.6 $m^3m^{-3}d^{-1}$ (Eden *et al.*, 1964). Reproduced with permission of the Chartered Institution of Water and Environmental Management.

that retention is probably irrelevant to percolating filter performance (Atkinson *et al.*, 1963; Mehta *et al.*, 1972).

Tracers can only measure their own retention and not necessarily the retention characteristics of waste liquids (Eden *et al.*, 1964). Originally, dyes, salt solutions or ammonium salt were used, but these often suffered from prolonged retention due to adsorption on to the biological film. In the late 1950s, however, experiments with radioactive tracers showed them to be ideal or a retention analysis. This was because: (a) only small quantities of such tracers are required, (b) of the ease and sensitivity of detection, and (c) of the negligible tendency of some of the radioactive substances to be adsorbed on to the biofilm (Eden and Melbourne, 1960). Since that time most research on retention analysis has involved the use of radioactive tracers, although Tariq (1975) published a method to calculate the mean time of retention by measuring the influent and drainage rates. A major disadvantage of Tariq's method is that the filter has to be shut down to measure the drainage characteristics, thus taking it out of operation for a number of hours, which would prove very difficult in smaller works. Due to the technical and financial problems of using radioactive isotopes, and also the possible environmental and health aspects of using them as tracers for the measurement of retention time, Gray (1981) used a simple technique using the traditional tracer sodium chloride and continuously monitored the conductivity of the final effluents to determine its recovery. He found the method of negligible cost, extremely quickly, and produced reproducible results, making it possible to include the regular measurement of retention time in the routine assessment of filter efficiency.

### 3.4.4 *Depth*

Most filters are built between 1.5 and 2.5 m in depth, with the majority approximately 1.8 m deep. It is traditional in some countries, such as Germany, to build them much deeper (3–5 m) but in practice, there seems little justification for constructing conventional filters that are deeper than 2.5 m. Depth does not appear to be a major factor in determining performance, although the total available surface area of the media is critical. Therefore, a shallow filter will give the same performance as a deeper filter of the same volume. Bruce and Merkens (1970) found no significant difference in performance between two filters of different depths containing plastic media, one 7.1 m deep the other 2.1 m, treating the same high volumetric loading.

Shallow filters (<1 m) have significant disadvantages. The risk of ponding and short-circuiting is far more serious in shallow filters. Without the extra depth to compensate, and as the depth increases, it becomes easier to obtain uniform distribution within the medium. Although shallow filters are structurally cheaper to construct, they require more land area, and the distribution cost will increase. In contrast, tall filters take up less area, and there is a significant increase in retention time with greater depth. However, increased pumping costs and increased structural costs make depth selection critical, and design engineers should avoid adding an extra 0.5 m or so to the height as an extra safety margin.

Gray (1980) studied the performance of low- and high-rate filters at various depths. In single-pass low-rate filters, three-quarters of all available BOD is removed in the top 900 mm of the bed, with between 45 and 50% removal occurring in the top 300 mm. The removal rate is affected by the biofilm accumulation, with large removals of BOD associated with relatively thin active films. This association continues until the biofilm becomes so thick that it begins to restrict the flow of wastewater through the filter, resulting in a short retention time due to channeling and a subsequent decrease in removal efficiency. At depths >900 mm, little BOD removal occurs (Table 3.12). More depth is utilized at high-rate loadings, illustrating the need for a greater surface area to oxidize the increased organic load. This is why most high-rate filters using plastic media are >5 m in height although they rarely exceed 7 m. Higher loading rates will increase the recirculation within filters using random plastic filter media, due to the high specific area of the medium, which reduces the possible increase in depth required. Suspended solids removal occur mainly in the top 900 mm at both low- and high-rate loadings, and is clearly linked with the adsorption capacity of the film. In the same filters, nitrification was restricted to depths of >900 mm, which increased to >1,500 mm, or even excluded altogether from the high-rate filters. Therefore, a minimum depth of 1 m is required to ensure adequate BOD and suspended solids removal in low-rate filters, although a minimum depth of 1.5 m is required if nitrification is also required. At higher loadings, a greater depth is required if enough surface area is to be made available for maximum BOD removal. If a random plastic medium is used, sufficient depth can also result in nitrification. Examples of preferred depths of percolating filters in various countries are given by Pike (1978).

### 3.4.5 *Biofilm*

Removal of organic matter occurs in four stages: (i) adsorption to the surface of the film, (ii) extracellular breakdown of adsorbed material, (iii) absorption by heterotrophic microorganisms, and (iv)

**Table 3.12.** The percentage removal of the total BOD in relation to depth in three experimental filters containing a variety of different media (Table 3.11), operated under low-rate conditions (1.68 $m^3m^{-3}d^{-1}$) (Gray, 1980).

| | Slag filter | | | | Mixed filter | | | | Plastic filter | | | |
|---|---|---|---|---|---|---|---|---|---|---|---|---|
| **Depth (mm)** | **0–300** | **300–900** | **900–1500** | **1500–1800** | **0–300** | **300–900** | **900–1500** | **1500–1800** | **0–300** | **300–900** | **900–1500** | **1500–1800** |
| Nov. 77 | 69.6 | 16.4 | 5.7 | 5.9 | 58.9 | 32.5 | 2.8 | 0.0 | 63.1 | 26.0 | 0.0 | 2.8 |
| Dec. 77 | 30.0 | 17.1 | 27.5 | 4.3 | 26.9 | 15.4 | 29. 7 | 0.0 | 30.0 | 27.1 | 16.3 | 0.0 |
| Jan. 78 | 20.6 | 61.6 | 1 2.2 | 0.0 | 40.0 | 46.1 | 0.0 | 3.4 | 47.5 | 45.4 | 3.7 | 1.5 |
| Feb. 78 | 32.7 | 21.6 | 45.6 | 0.0 | 43.7 | 46.1 | 0.0 | 4.5 | 66.9 | 31.9 | 0.0 | 0.8 |
| Mar. 78 | 40.8 | 33.8 | 13.3 | 6.5 | 53.2 | 19.4 | 22.5 | 0.3 | 58.7 | 22.5 | 3.4 | 6.6 |
| Apr. 78 | — | — | — | — | — | — | — | — | — | — | — | — |
| May. 78 | 59.0 | 27.9 | 0.0 | 2.7 | 62.0 | 19.3 | 14.4 | 0.0 | 56.9 | 24.6 | 6.9 | 2.8 |
| Jun. 78 | 38.8 | 37.3 | 0.0 | 12.4 | 34.4 | 50.3 | 0.6 | 4.5 | 68.3 | 0.4 | 10.1 | 3.9 |
| Jul. 78 | 58.6 | 25.7 | 3.7 | 7.5 | 57.2 | 26.8 | 3.0 | 6.8 | 45.5 | 27.1 | 1 9.0 | 0.0 |
| Aug. 78 | 73.1 | 15.9 | 0.0 | 1.9 | 48.1 | 39.5 | 3.6 | 1.9 | 45.7 | 37.9 | 9.9 | 0.0 |
| Sep. 78 | 44.2 | 20.3 | 17.9 | 7.9 | 68.4 | 6.8 | 12.3 | 0.0 | 42.2 | 27.3 | 10.2 | 8.1 |
| Mean | 46.74 | 27.76 | 12.59 | 4.91 | 49.28 | 30.22 | 8.89 | 2.14 | 52.48 | 27.02 | 7.95 | 2.65 |
| S.d. | 17.56 | 13.96 | 14.69 | 3.91 | 13.08 | 14.96 | 10.47 | 2.50 | 12.29 | 11.62 | 6.43 | 2.84 |

mineralization. As the first step in purification is adsorption, any limitation on the availability of adsorption sites will reduce purification. The removal efficiency of both BOD and suspended solids in wastewater is linked to a thin actively growing film. The grazing fauna is very important in helping to maintain an actively growing biofilm, thus ensuring maximum rates of adsorption. The rate of adsorption is independent of temperature, whereas the rate of oxidation is temperature-dependent.

The factor that most influences retention time, assuming the hydraulic flow remains constant is the biofilm accumulation. Eden *et al.* (1964) demonstrated that retention time increased directly with biofilm accumulation up to optimum weight, after which retention characteristics changed due to excessive biofilm (Sec. 3.4.3). Heavy weights of biofilm alter the flow pattern. This results in the storage of effluent as localized reservoirs within filters, channeling of the wastewater, and eventual ponding, which results in a loss of performance. The biofilm accumulation also follows a seasonal pattern. It becomes thicker during the winter months due to reduced bioxidation and reduced grazing at the lower temperatures (Gray, 1983a) (Fig. 3.9). Therefore, the most important influence on filter efficiency is the volume and growth rate of film, which not only controls the retention time, but also the rate of adsorption of suspended and dissolved nutrients, the rate of bioxidation, and the sludge production. In operational terms, a knowledge of the biofilm accumulation within a filter is required if maximum performance efficiency is to be maintained (Honda and Matsumoto, 1983). Ideally, the quantity and quality of active biomass within the filter should be controlled as in the activated sludge process, which can only be achieved in percolating filters by having a greater fundamental understanding of the kinetics of biofilm growth and accurate methods of determining biofilm accumulation. A number of modifications of the percolating filter process have been developed such as random and modular plastics media, recirculation, ADF, control of the frequency of dosing, and modifications to the distribution system (Sec. 3.3). One major advantage of these modifications is that they attempt to achieve optimal biofilm growth. However, even with such modifications, it is desirable to have some indication of the biofilm accumulation within the filter, especially as depth studies have shown that the surface biofilm accumulation rarely resembles the biofilm accumulated within the filter (Gray, 1983b). With modifications such as ADF and recirculation, the stage of biofilm development is an important factor in deciding whether the filter series should be changed or the ratio of the returned effluent modified.

Although a number of possible methods of estimating the degree of biofilm accumulation are available, for example, the adsorption of dyes (Smith and Coackley, 1983), or comparative retention time analysis, the relationship between these parameters and biofilm weight is far from clear. The information obtained in this way is also rather imprecise, with one value obtained for the condition of the entire filter, but no estimation of biofilm accumulation at specific depths or areas of the filter is possible. Three approaches to biofilm measurement are generally used. In some industrial filter installations employing plastic medium, it is possible to measure the total weight of the filter plus the accumulated biofilm by sensors built into the base of the reactor to give a total biofilm weight. A better idea of where biofilm is accumulating within the filter is achieved by either using sampling tubes containing baskets of the medium that can be removed and the biofilm estimated by a number of gravimetric methods, or an indirect assessment method where the biofilm is measured within the filter without disturbing the medium by using a neutron modulation technique (Sec. 4.5).

The biofilm is comprised mainly of water between 92 and 97% in plastic media filters and 95–98% in mineral media filters (Gray, 1980). This allows biofilm accumulation to be estimated in filters by measuring the abundance of hydrogen atoms present in the water molecules using neutron modulation. Fast neutrons emitted from a radioactive isotope (beryllium-americium) form a cloud of slow neutrons on collision with hydrogen atoms. By using a slow neutron detector the hydrogen atoms in any water present in the filter can be measured (Bell, 1973) (Fig. 3.17). There are several commercial neutron probes available, although the most widely used probe in wastewater studies is the Wallingford Soil Moisture Probe. The method was

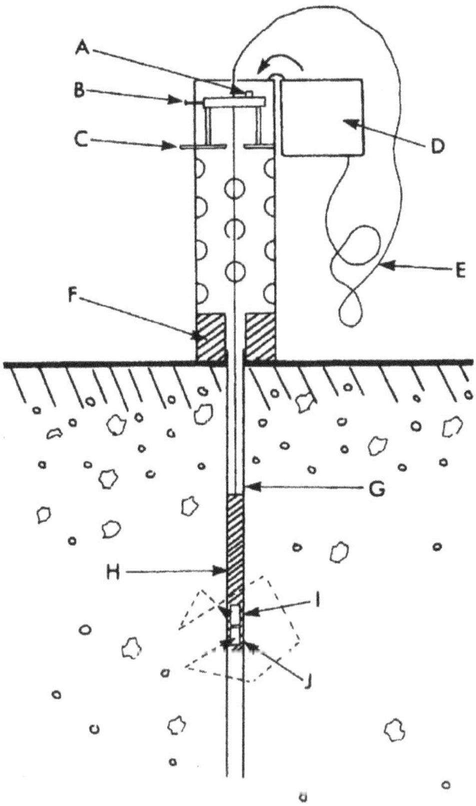


**Fig. 3.17.** Diagram showing the neutron probe in use: Depth indicator (A), cable clamp (B), lock (C), rate meter (D), cable (E), plastic shield (F) containing isotope, filter access tube (G), probe (H), slow neutron detector (I) and neutron source (J).

developed at the Centre for Ecology and Hydrology at Wallingford in the UK, which has developed a non-invasive method using natural background neutrons, thus eliminating the need for an artificial source. However, this new technology has not been applied to filter systems. The probe is lowered into the filter via a 50 mm diameter aluminum access tube, which is sealed when not in use to keep it dry. The probe has to be calibrated for each medium tested using a 210-liter drum containing the medium over a range of moisture levels (Fig. 3.18). Specific and significantly different calibration curves are obtained for each medium used, with the difference between the

(a) (b)

**Fig. 3.18.** The neutron probe in use. (a) Probe being calibrated in the laboratory. The probe is lowered into an aluminum access tube embedded in the test medium. (b) In use measuring biofilm accumulation within a pilot-scale filter.

slope of the calibration curves for different media because of the absorption of thermal neutrons by the media (Harvey *et al.*, 1963). Therefore, the appropriate calibration curve must be used when converting the probe readings into the percentage saturation of the voids. The use of neutron modulation in the estimation of biofilm accumulation has been fully described by Harvey *et al.* (1963) and Gray (1984).

Gray and Learner (1984) compared direct gravimetrc measurements of biofilm weight with the neutron scattering technique. Four gravimetric methods of biofilm accumulation were used — total biofilm weight, total dry solids, volatile solids, and percentage settlement of solids, which incorporated both weight and volume measurements. Total biofilm was the weight increase of the sampling basket and the medium due to accumulated film. The wet weight of the clean medium in each sampling basket was measured at the start of the investigation. Once a month, each basket was removed from the sampling tube and the excess liquid was drained off for five minutes and then weighed, with the increase in weight being due to biofilm growth. Total dry solids were measured by evaporating a 50 $cm^3$ subsample of the film (removed from a 1-liter sample of filter medium) in a weighed evaporating dish to dryness at 105°C for 24 hours. The dish was reweighed after cooling to room temperature. The quantity of volatile solids was determined by removing the organic fraction of the dried sample by burning it in a muffle furnace at 500°C for 1 hour before the dish was reweighed after cooling to room temperature. An assessment of the volume of solids present was made using the Imhoff cone method. A 250 $cm^3$ subsample of the removed biofilm was allowed to settle for 45 minutes in the Imhoff cone, which was then gently twisted to remove any debris adhering to the glass sides. The quantity of solids that settled after an hour was recorded as the percentage settlement. Although any large lumbricid worms were removed before settlement, all the samples contained large numbers of invertebrates, and these were included in this assessment of biofilm accumulation. All the gravimetric estimations were made at monthly intervals at six depths within each filter, corresponding with the depths at which the biofilm was estimated by the neutron scattering technique. Before neutron scattering can be used, all the excess liquid has to be drained out of the filters once the distribution system has been shut off. Readings were taken at 150 mm intervals throughout the depth of the filters, and each value taken was a mean of five readings, each of which was a mean count per second integrated over a 16-second sample period. Replicate access tubes were used to allow the biofilm accumulation to be studied from either side of each filter and the mean of the two average readings at each depth was taken to represent the mean percentage saturation of the voids. Neutron probe readings were always taken within 24 hours of the gravimetric sampling. The results of the comparative study by Gray and Learner (Table 3.13) showed good correlations ($P < 0.001$) between all the methods tested in low-rate conditions (1.68 $m^3m^{-3}d^{-1}$ and 0.28 kg BOD $m^{-3}d^{-1}$), although the neutron scattering results were not significantly correlated ($P > 0.05$) with any of the gravimetric methods at the higher loading (3.37 $m^3m^{-3}d^{-1}$ and 0.63 kg BOD $m^{-3}d^{-1}$). They concluded that although neutron scattering provided a rapid and sensitive measure of hydrogen atoms in the filter, the results

**Table 3.13.** Summary of the range of biofilm accumulations measured by the neutron scattering technique (NP), and also as the total film (TF), volatile solids (VS), total solids (TS), and percentage settlement (PS) in mineral and plastics media at two different loading rates (Gray and Learner, 1984). Reproduced with permission of Elsevier Publishing.

| | Mineral medium | | | | | Plastic medium | | | | |
|---|---|---|---|---|---|---|---|---|---|---|
| | NP | TF | VS | TS | PS | NP | TF | VS | TS | PS |
| | (% sat) | (kg m$^{-3}$) | | | (%) | (% sat) | (kg m$^{-3}$) | | | (%) |
| | *Low loading rate* (1.68 m$^3$ m$^{-3}$d$^{-1}$) | | | | | | | | | |
| Mean | 23.0 | 257 | 7.25 | 10.08 | 16.0 | 3.5 | 104 | 5.01 | 6.60 | 11.0 |
| Minimum | 11.3 | 96 | 1.04 | 1.48 | 2.4 | 1.1 | 28 | 0.36 | 0.44 | 0.6 |
| Maximum | 39.0 | 616 | 23.08 | 31.44 | 60.0 | 15.6 | 315 | 16.44 | 21.28 | 37.2 |
| Range | 27.7 | 520 | 22.04 | 29.96 | 57.6 | 14.5 | 287 | 16.08 | 20.84 | 36.6 |
| Skewness | 0.09 | 0.75 | 0.99 | 0.96 | 1.37 | 2.22 | 0.72 | 0.69 | 0.60 | 0.69 |
| *n* | 66 | 54 | 66 | 66 | 66 | 66 | 54 | 66 | 66 | 66 |
| | *High loading rate* (3.37 m$^3$ m$^{-3}$d$^{-1}$) | | | | | | | | | |
| Mean | 26.5 | 216 | 5.91 | 7.68 | 11.1 | 4.5 | 173 | 8.34 | 10.91 | 18.8 |
| Minimum | 16.9 | 8.3 | 1.32 | 1.88 | 2.2 | 1.3 | 34 | 1.20 | 1.36 | 1.4 |
| Maximum | 40.2 | 494 | 23.68 | 32.56 | 48.0 | 12. 2 | 474 | 25.08 | 33.92 | 78.4 |
| Range | 23.3 | 411 | 22.36 | 30.68 | 45.8 | 10.9 | 440 | 23.88 | 32.56 | 77.0 |
| Skewness | 0.25 | 1.31 | 2.22 | 2.37 | 1.85 | 1.06 | 0.91 | 1.20 | 1.28 | 1.62 |
| *n* | 12 | 72 | 72 | 72 | 72 | 12 | 72 | 72 | 72 | 72 |

expressed as the percentage saturation of the voids are not directly transferable to biofilm weights and should be treated separately and not as a true measure of biofilm accumulation. Neither could the technique provide qualitative information about the biofilm that was obtained by visually inspecting the exposed cores of biofilm covered medium. It does, however, clearly indicate to the operator areas of high or low water saturation. It, therefore, appears to be ideal for use in the routine operational control of filter modifications, such as ADF and recirculation, and also in predicting the onset of ponding, especially below the surface of the medium. However, accurate measurement of biofilm weight and a detailed analysis of the condition and composition of the biofilm are only possible by using gravimetric analyses.

Film accumulation data is therefore useful in indicating where biomass is building up within the filter and can inform the operator if potential ponding problems exist, if the filter is being over or underloaded, or if biofilm development is inhibited. There are a number of ways of visually interpreting such data. Traditionally, kite diagrams have been used to illustrate biofilm accumulation at various depths, with a separate kite plot for each month or sampling period (Fig. 3.19). A development of this is the horizontal histogram, which, when turned on its side, gives a three-dimensional view of the changes over the sampling period of vertical biofilm accumulation within the filter bed. Contour mapping of biofilm data allows biofilm concentration contours to be plotted, with depth and time being the only axes. This approach allows interpolation between the available data points,

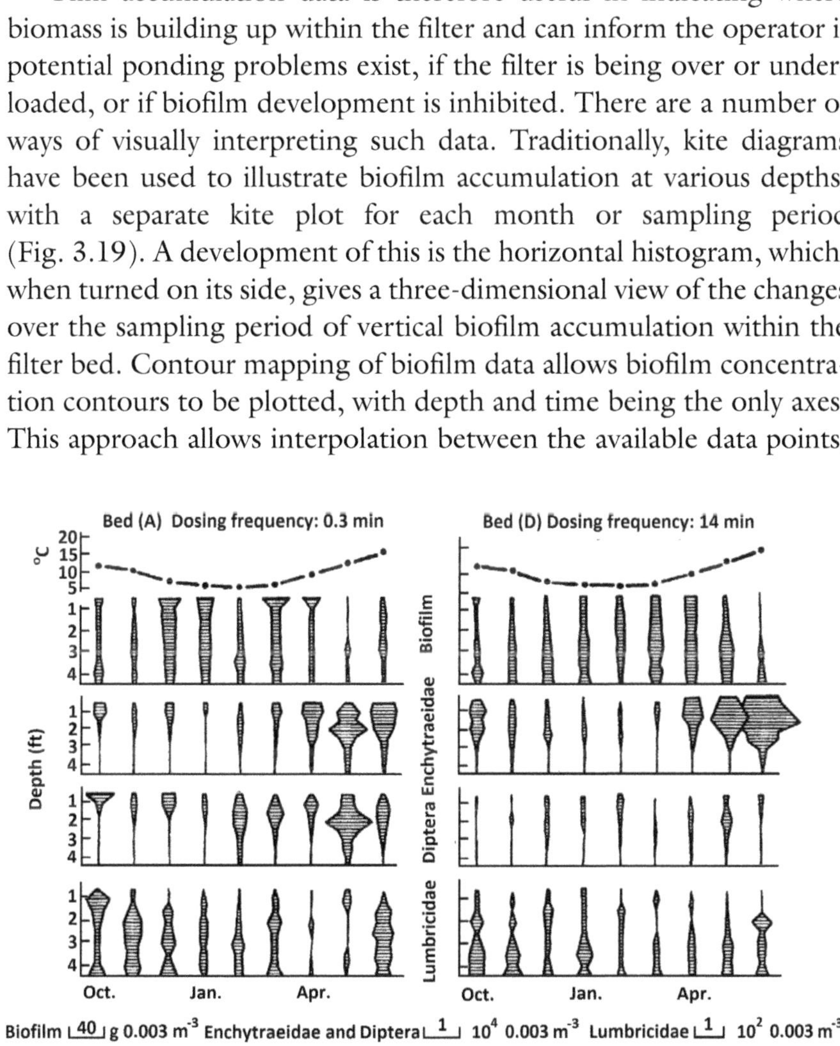


**Fig. 3.19.** Kite diagram showing seasonal changes in the abundance and vertical distribution of biofilm and grazing fauna in a high-frequency dosed filter (A) and a low-frequency dosed filter (D) (Hawkes and Shephard, 1972). Reproduced with permission of Elsevier Publishing.

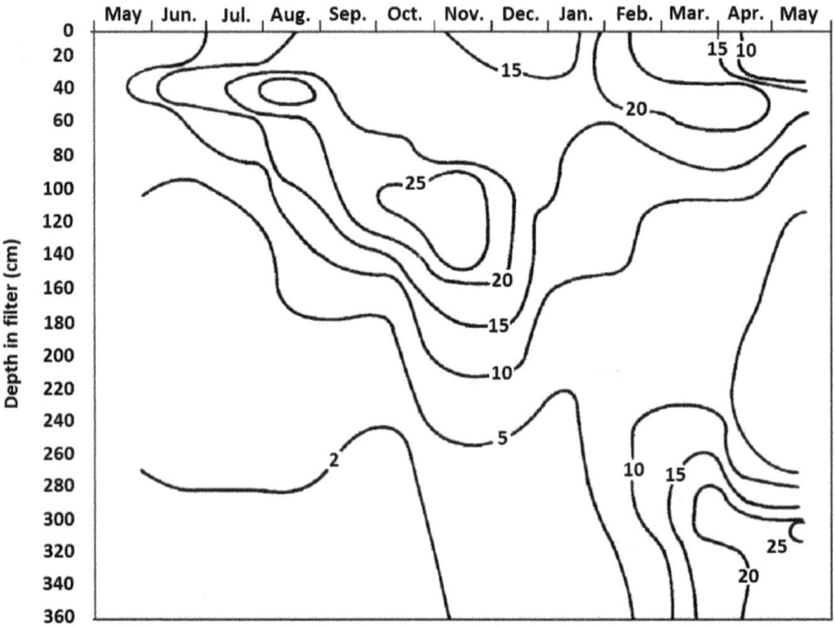


**Fig. 3.20.** Contour mapping of the biofilm accumulation (as percentage saturation of the voidage) in a random plastic filter medium (Flocor RC™) receiving finely screened sewage (Hoyland and Roland, 1984). Reproduced with permission of WRc PLC.

normally by simple linear regression. By selecting the number of concentration contour levels to be plotted, either simple or very detailed views of seasonal biofilm accumulation within the filter can be obtained (Fig. 3.20).

### 3.4.6 *Ventilation*

Oxygen is supplied to the aerobic microorganisms comprising the biofilm by absorption from the air as it passes through the filter by natural ventilation. When ventilation ports are blocked or partially restricted, the performance of the filter is seriously affected; therefore, good ventilation is vital for the successful operation of percolating filters. Ventilation currents are caused by the temperature difference between the air outside and the liquid and biofilm inside the filter, and the resultant flow of air may be in either direction. In winter, the air entering the filter will be warmed by the wastewater passing through the bed, causing a net upward flow. In summer, the opposite may occur with the air possibly being cooled in the filter, resulting in a downward flow of air. The rate of flow of air through the filter is directly proportional to the difference in the temperature between the outside air and the wastewater, with a difference of 10°C inducing a rate of flow of air of about 20 $m^3m^{-2}d^{-1}$. Other factors also affect ventilation, such as wind currents and humidity. The humidity of the air inside the filter will increase as it passes through, and as an increase in humidity reduces the density of air, this encourages an upward flow. While the oxygen concentration of air in the interstices within the filter remains very close to that of normal air (20.9% v/v), reduction of the oxygen concentration is possible during periods of high carbonaceous oxidation. It is unlikely that normal heterotrophic activity will be affected by low oxygen concentrations, with no reduction in BOD removal at oxygen concentrations of 2% v/v, although nitrification is far more sensitive to any oxygen limitation.

The rate of oxygen uptake in percolating filters varies from 0.273 kg $O_2$ $m^{-3}d^{-1}$ in low-rate filters, where 0.187 kg $m^{-3}d^{-1}$ is required for heterotrophic oxidation and 0.086 kg $m^{-3}d^{-1}$ for nitrification (Montgomery and Borne, 1966), to 1.490 kg $m^{-3}d^{-1}$ for high-rate filters without nitrification (Bruce and Merkens, 1970). The rate of oxygen utilization depends on the specific surface area of the media on which the biofilm develops, thus, for a medium with a specific surface area of 200 $m^2m^{-3}$, the potential uptake rate of oxygen could be as high as 1.9 kg $m^{-3}d^{-1}$. The rate of oxygen utilization is fairly uniform throughout low-rate filters with demand from heterotrophs in the upper layer declining with depth, and the demand from nitrifiers increasing with depth. The rates at which oxygen and organic substrate diffuse into the biofilm depend on their relative concentrations in the liquid phase covering the film. For example, when the substrate concentration is high, then the concentration gradient is larger, thus causing more rapid diffusion to a much greater depth. The depth of penetration into the biofilm is approximately proportional to the substrate concentration (Atkinson and Fowler, 1974). The depth of penetration of organic nutrients into the biofilm doubled from 0.06 to 0.15 mm when the substrate concentration was increased from 10 to 500 g of COD $m^{-3}$. In practice, the maximum substrate concentration is found in the upper 900 mm of filters, where the maximum biofilm accumulation is also recorded. The rate of diffusion in this area of the filter may exceed the rate at which oxygen can diffuse into the biofilm for aerobic metabolism, resulting in incomplete oxidation. Therefore, in the top section of the filter, the oxidation rate will be oxygen-limited, whereas, with depth, the rate of oxidation will decrease due to a reduction in the substrate concentration. At lower substrate concentrations, the organic nutrients can only penetrate a short distance into the biofilm before being consumed, so the rate of oxidation is substrate limited here (Fig. 3.21). The remaining thickness of the biofilm is important as it acts as a buffer against environmental and operational changes. For example, if the temperature falls, the nutrients would penetrate further into the biofilm before being consumed, thereby using more of the microbial population. This has been modeled by Harris and Hansford (1976), who suggested that oxygen limitation occurs when the substrate concentration exceeded 400 g COD $m^{-3}$. Williamson and McCarty (1976)

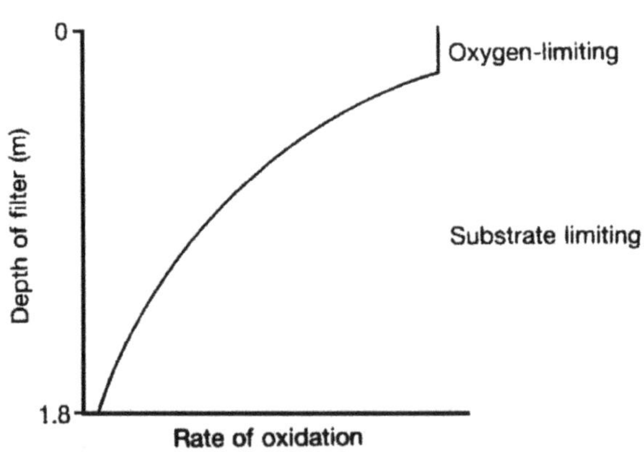


**Fig. 3.21.** The rate of oxidation in relation to the depth of a percolating filter indicating where oxygen and substrate concentrations are major limiting factors.

have also modeled this phenomenon and suggested that oxygen becomes the limiting factor in biofilm when:

$$CK_c < SK_s,$$

where $\boldsymbol{C}$ and $\boldsymbol{S}$ are the concentrations of oxygen and BOD in the biofilm, respectively, and $\boldsymbol{K}_c$ and $\boldsymbol{K}_s$ are their respective saturation coefficients.

Numerous methods have been employed to improve ventilation including leaving gaps between stonework, using special air bricks, providing ventilation tubes through the medium, and even using fans and pumps to force air through the bed. However, it appears that such systems are unnecessary as normal ventilation systems using air vents attached to the underdrains will provide adequate amounts of oxygen for biological activity, unless the voids are excessively clogged by the accumulated film.

Normally odor is not a problem associated with low-rate filtration so long as the plant is in good working order. Occasionally, when the air and influent temperature are the same, especially when there is little air movement, ventilation may be suppressed for short periods resulting in stale air being produced when air movement recommences. Likewise, when the filter is shut down for any length of time during warm weather, odors will result due to a loss of aerobic conditions within the bed. Odors can be produced where: (i) the filter treats septic wastewater, (ii) the filter treats industrial wastewater containing high concentrations of sulfate, and (iii) the wastewater contains a significant portion of seawater. In all three cases hydrogen sulfides and other associated malodors may be produced. If the filter ponds or becomes choked with biofilm so that insufficient ventilation results in occurring anaerobic areas, malodors will result (Gray, 2004).

In contrast, high-rate filters can cause odor problems as they act as very efficient vapor strippers, especially of ammonia. This is a particular problem with industrial installations, such as pharmaceutical factories, where solvents are widely employed in the manufacturing process. To control this, industrial high-rate filters are normally sealed, thus requiring a forced ventilation system to be used. This has the advantage that the temperature within the filter and the oxygen availability can also be controlled to ensure maximum oxidation rates. Slow air movement through deep filters operating at high loading rates may also lead to odors and loss of performance, thus requiring the use of forced ventilation systems (Suschka, 1987). Some odor production is inevitable from the high-rate filters due to the nature of the plastic medium. The biofilm rapidly develops and regularly sloughs away from the surface of the smooth medium, usually when the inner layer of the biofilm becomes anaerobic. This results in the release of malodors whenever sloughing occurs.

### 3.4.7 *Frequency of Dosing*

Influent wastewater is distributed on to the surface of percolating filters via sprayers or nozzles mounted on either a moving distributor or a system of static distributor pipes. Although the same loading rate is applied to both types of filters, the former distributor arrangement doses each specific area intermittently, whereas the latter doses the entire surface of the filter continuously. It is general practice to use moving distributors on all filters except very small installations where a fixed system of distributor pipes is used, and intermittent dosing is achieved by using balancing tanks that are emptied at the required interval, or tipping-trough systems. Filters containing plastic media that have a high voidage are generally loaded continuously from a fixed distribution system, as this gives a better distribution of wastewater within the filter. Under low-rate conditions, the best performance is obtained using large doses at infrequent intervals, i.e., low speeds of rotation. In this way, each segment of the filter bed is subjected to a surge of liquid, followed by a long period without any flow before the next dose. The slower the speed of rotation the greater the surge of wastewater and the longer the intervening period between doses. In theory, the long period between dosing ensures that the microorganisms comprising the biofilm are in a substrate-limited condition so that they are in the endogenous respiration growth phase, i.e., a low f/m condition, which limits biofilm accumulation

(Gray, 2004). This has been especially effective in preventing fungal growth in winter. The surge of wastewater ensures a uniform distribution of organic nutrients within the bed, extending the depth of heterotrophic activity and encouraging a more even distribution of the film. The mechanical scouring effect also limits biofilm development, but this is of minor significance compared with the limitation of nutrients. By lowering the frequency of dosing the retention time is also reduced. Therefore, dosing frequency must be a compromise between maximizing the retention time so that sufficient wastewater-film contact time is maintained, in order that maximum adsorption of the organic nutrients can occur, and excessive biofilm accumulation controlled. High-frequency dosing approaches the conditions found with continuous dosing with a large accumulation of microbial biofilm at the surface. High-frequency dosing is ideal for random plastic medium or media with a high voidage that can support a high biofilm accumulation in the top 900 mm. It is also suitable for mineral media filters that have been uprated by replacing the surface layer of medium with a random plastic medium (Gray and Learner, 1984).

The optimum frequency of dosing for low-rate filters is between 15–30-minute intervals (Tomlinson and Hall, 1950). This prevents ponding in the winter, reduces the population of the fly *Sylvicola fenestralis* emerging during the warmer months, and encourages the enchytraeid worms to become the dominant grazer, replacing the dipteran larvae (Hawkes, 1955) (Sec. 4.4). However, the optimum frequency of dosing for a specific filter varies with the operating conditions, the type of medium, and the nature of the wastewater. The optimum frequency of dosing will also change seasonally, with it being much lower in the winter in order to control excessive biofilm accumulation when compared to the summer when other factors control the biofilm (Bruce and Hawkes, 1983).

To control the frequency of dosing, the rotating arms need to be propelled using a motor-driven rotor or a peripheral wheel driven by the wastewater flow using a water-wheel arrangement. Although the speed of rotary arms driven by the reaction of the spray jets alone can be controlled to a certain extent by altering the angle of the jets, reaction driven rotation requires a minimum head of 0.5–1.0 m and is severely affected by changes in the wind direction. Some form of mechanical drive is required, therefore, for accurate periodic dosing of filters.

## References

Agra, C.A. (2009) Tratamento de águas residuárias domésticas em reatores de biomassa dispersa e biomassa aderida. Dissertação de mestrado. Universidade Federal da Paraíba/Universidade Estaduala Paraíba. Campina Grande, Brazil.

Ali, I., Khan, Z.M, Sultan, M., Mahmood, M.H., Farid, H.U., Ali, M. and Nasir, A. (2016) Experimental study on maize cob trickling filter-based wastewater treatment system: Design, development, and performance evaluation. *Polish Journal of Environmental Studies*, **25**, (6), 2265–2273.

Alvarez-Hornos, F.J., Izquierdo, M., Martinez-Soria, V., Oenya-Roja, J.M., Sempere, F. and Gaballdon, C. (2011) Influence of ground tire rubber on the transient loading response of a peat biofilter. *Journal of Environmental Management*, **92**, (8), 1978–1985.

Anderson, G.K. (1977) Farmland food wastes. In *Treatment of Industrial Effluents* (eds. A.G. Callely, C.F. Forster and D.A. Stafford), Hodder and Stoughton, London, pp. 245–257.

Anon. (1979) Advanced wastewater treatment: A low-risk proving-ground for experiment and invention. *World Water*, **6**, 29–41.

Aslam, M.M.A., Khan, Z.M., Sultan, M., Niaz, Y., Mahmood, M.H., Shoaib, M., Shakoor, A. and Ahmad, M. (2017) Performance evaluation of trickling filter-based wastewater treatment system utilizing cotton sticks as filter medium. *Polish Journal of Environmental Studies*, **26**, (5), 1955–1962.

Atkinson, B., Busch, A.W. and Dawkins, G.S. (1963). Recirculation reaction kinetics and effluent quality in a trickling filter flow model. *Journal of the Water Pollution Control Federation*, **35**, 1307–1317.

Atkinson, B. and Fowler, H.W. (1974) The significance of microbial film in fermenters. *Advances in Biochemical Engineering*, **3**, 221–277.

Balan, C., Bulai, P., Bilba, D. and Macoveanu, M. (2010) Sphagnum moss peat: A green and economical sorbent for removal of heavy metals (Cd and Cr) from wastewaters. *Environmental Engineering and Management Journal*, **9**, (4), 469–477.

Barnard, J.L. and Steichen, M.T. (2006) Where is biological nutrient removal going now? *Water Science and Technology*, **53**, (3), 155–164.

Bayley, R.W. and Downing, A.L. (1963) Temperature relationships in percolating filters. *Institution of Public Health Engineers Journal*, **62**, 303–312.

Bell, J.P. (1973) *Neutron Probe Practice.* Report No. 19. Institute of Hydrology, National Environment Research Council.

Besselievre, E.B. and Schwartz, M. (1976) *The Treatment of Industrial Wastes, 2nd ed.* McGraw-Hill.

Bhatnagar, A. and Sillanpaa, M. (2010) Utilization of agro-industrial and municipal waste materials as potential adsorbents for water treatment: A review. *Chemical Engineering Journal*, **157**, (2–3), 277–296.

Boltz, J.P., Morgenroth, E., DeBarbadillo, C., Dempsey, M.J., Ghylin, T., Harrison, J., McQuarrie, J.P. and Nerenberg, R. (2010) Biofilm reactor technology and design. In *Design of Municipal Wastewater Treatment Plants, 5th ed.*, Water Environment Federation, the American Society of Civil Engineers, and the Environmental and Water Resources Institute, Alexandria, Virginia, US, pp. 13.1–13.237.

Bruce, A.M. (1968) The significance of particle shape in relation to percolating filter media. *Journal of the British Granite Whinstone Federation*, **8**, (2), 1–15.

Bruce, A.M. (1969) Percolating filters. *Process Biochemistry*, **4**, (4), 19–23.

Bruce, A.M. and Hawkes, H.A. (1983) Biological filters. In *Ecological Aspects of Used-Water Treatment, Vol. 3. The Processes and their Ecology* (eds. C.R. Curds and H.A. Hawkes). Academic Press, London, pp. 1–111.

Bruce, A.M. and Merkens, J.C. (1970) Recent studies of high-rate biological filtration. *Water Pollution Control*, **69**, 113–148.

Bruce, A.M. and Merkens, J.C. (1973) Further studies of partial treatment of sewage by high-rate biological filtration. *Water Pollution Control*, **72**, 499–523.

Bruce, A.M., Merkens, J.C. and Haynes, B.A.O. (1975) Pilot-scale studies on the treatment of domestic sewage by two-stage biological filtration with special reference to nitrification. *Water Pollution Control*, **74**, 80–100.

Bruce, A.M., Truesdale, G.A. and Mann, H.T. (1967) The comparative behaviour of replicate pilot-scale percolating filters. *Journal of the Institution of Public Health Engineers*, **66**, 151–175.

BSI (1948) Specification for media for biological filters, BS 1438. British Standards Institution, London.

BSI (1971) Specification for media for biological percolating filters, BS 1438:1971. British Standards Institution, London.

BSI (1980) Media for biological percolating filters, BS1438:1971 as amended by AMD 998 and AMD 3454. British Standards Institution, London.

BSI (1986) Specification for test sieves, BS 410:1986. British Standards Institution, London.

BSI (1989) Test sieving. Methods using test sieves of woven wire cloth and perforated metal plate, BS 1796-1:1989 ISO 2591-1 1988. British Standards Institution, London.

BSI (2004) Aggregates. Single-sized aggregates for use as a medium in biological percolating filters, BS 1438:2004. British Standards Institution, London.

BSI ISO (2013) Test sieves. Technical requirements and testing. Test sieves of perforated metal plate, BS ISO 3310-2:2013. British Standards Institution, London.

Calley, A.G., Forster, C.F. and Stafford, D.A. (1977) *Treatment of Industrial Effluents*. Hodder and Stoughton, London.

Chipperfield, P.N.J. (1967) Performance of plastic filter media in industrial and domestic waste treatment. *Journal of the Water Pollution Control Federation*, **39**, 1860–1874.

CIWEM (2000) *Biological Filtration and Other Fixed Film Processes*. The Chartered Institution of Water and Environmental Management, London.

Clark, S. and Pitt, R. (2009) Storm-water filter media pollutant retention under aerobic versus anaerobic conditions. *Journal of Environmental Engineering*, **135**, (5), 367–371.

Cook, E.E. and Herning, L.P. (1978) Shock load attenuation trickling filter. *Journal of the Environmental Engineering Division*, **104**, (EE3), 461–469.

Cook, E.E. and Katzberger, S.M. (1977) Effect of residence time on fixed film reactor performance. *Journal of the Water Pollution Control Federation*, **49**, 1889–1895.

Cooke, W.B. and Hirsche, A. (1958) Continuous sampling of trickling. *Sewage and Industrial Wastes*, **30**, 138–155.

Corbett, J. (1902) Some purification experiments. *Journal of the Sanitation Institute,* **23**, 601–608.

Corley, M., Rodgers, M, Mulqueen, J. and Clifford, E. (2006) The performance of fibrous peat biofilters in treating domestic strength wastewater. *Journal of Environmental Science and Health Part A — Toxic/Hazardous Substances and Environmental Engineering*, **41**, 811–824.

Craft, T.F. and Ingols, R.S. (1973) Flow through time in trickling filters. *Water and Sewage Works,* **120**, (1), 78–79.

Curds, C.R. and Hawkes, H.A. (eds.) (1975) *Ecological Aspects of Used-Water Treatment, Vol. 1. The Organisms and their Ecology.* Academic Press, London.

Dudley, J.L., Arthurs, W. and Hall, T.J. (2001) A comparison of methods used to estimate river rock surface areas. *Journal of Freshwater Ecology,* **16**, 257–261.

Earle, M. (1986) A preliminary survey of percolating filters in Ireland. M.Sc. Thesis. Trinity College, University of Dublin, Dublin.

Earle, M. and Gray, N.F. (1987) Check list of Irish wastewater treatment plants: Percolating filters. Technical Report 2. Water Technology Research, Dublin.

Eckenfelder, W.W. (1980). *Principles of Water Quality Management.* CBI Publishing Co. Inc., Boston.

Eden, G.E., Brendish, K. and Harvey, B.R. (1964) Measurement and significance of retention in percolating filters. *Journal and Proceedings of the Institute of Sewage Purification,* **6**, 513–525.

Eden, G.E. and Melbourne, K.V. (1960) Radioactive tracers for measuring the periods of retention in percolating filters. *International Journal of Applied Radiation and Isotopes,* **8**, 172–178.

Escritt, L.B. (1965) *Sewerage and Sewage Disposal.* C.R. Brooke, London.

Farnham, R.S. (1974) Use of organic soils for wastewater filtration. *Soil Science Society of America. Special Publication,* **6**, 111–118.

Forster, C.F. (1977) Bio-oxidation. In *Treatment of Industrial Effluents* (eds. A.G. Cullely, C.F. Forster and D.A. Stafford). Hodder and Stoughton, London, pp. 65–87.

Garzon-Zuniga, M.A. and Buelna, G. (2011) Treatment of wastewater from a school in a decentralized filtration system by percolation over organic packing media. *Water Science and Technology,* **64**, 1169–1177.

Gebert, W. and Wilderer, P.A. (2000) Heating up tickling filters to tackle cold weather conditions. *Water Science and Technology,* **41**, (1), 163–166.

Gray, N.F. (1980) The comparative ecology of high-rate plastic, conventional mineral and mixed plastic/mineral media in the treatment of domestic sewage in percolating filters. Vol. I. Ph.D. Thesis. Council for National Academic Awards, London.

Gray, N.F. (1981) Simple retention time analysis in percolating filters using conductivity. *Effluent and Water Treatment Journal,* **21**, 345–347.

Gray, N.F. (1983a) Ponding of a random plastic filter medium due to the fungus *Subbaromyces splendens* Hesseltine in the treatment of sewage. *Water Research*, **17**, 1295–1302.

Gray, N.F. (1983b) The significance of the flow pattern on retention analysis in percolating filters. *Irish Journal of Environmental Science*, **2**, 293–297.

Gray, N.F. (1984) Biological film control in percolating filters using neutron scattering. *Effluent and Water Treatment Journal*, **24**, 203–205.

Gray, N.F. (2004) *Biology of Wastewater Treatment.* Imperial College Press, London.

Gray, N.F. (2010) Failure of plastic media biofilters due to ponding caused by the fungus *Subbaromyces splendens* Hesseltine, *Water Today*, **1**, (5), 86–90.

Gray, N.F. and Coady, J. (2015) Evaluation of full-scale vertical peat biofilters for treatment of septic tank effluents, *International Journal of Environmental Studies*, **72**, (1), 154–165.

Gray, N.F. and Learner, M.A. (1983) A pilot-scale percolating filter for use in sewage treatment studies. *Water Research*, **17**, 249–253.

Gray, N.F. and Learner, M.A. (1984) Estimation of film accumulation in percolating filters: A comparison of methods. *Water Research*, **18**, (12), 1509–1513.

Harris, N.P. and Hansford, G.S. (1976) A study of substrate removal in a microbial film reactor. *Water Research*, **10**, 935–943.

Harrison, J.R. and Daigger, G.T. (1987) A comparison of trickling filter media. *Journal of the Water Pollution Federation*, **59**, 679–685.

Harvey, B.R., Eden, G.E. and Mitchell, N.T. (1963) Neutron scattering: A technique for the direct determination of the amount of biological film in a percolating filter. *Journal and Proceedings of the Institute of Sewage Purification*, (5), 495–506.

Hawkes, H.A. (1955) The effect of periodicity of dosing on the amount of film and the numbers of insects and worms in alternating filters at Minworth. *Journal and Proceedings of the Institute of Sewage Purification*, (1), 48–58.

Hawkes, H.A. (1957) Film accumulation and grazing activity in the sewage filters at Birmingham. *Journal and Proceedings of the Institute Sewage Purification*, (1), 88–110.

Hawkes, H.A. (1963) *The Ecology of Wastewater Treatment.* Pergamon, Oxford.

Hawkes, H.A. and Jenkins, S.H. (1955) Comparison of four grades of sewage percolating filter media in relation to purification, film accumulation and fauna. *Journal and Proceedings of the Institute of Sewage Purification*, (3), 352–357.

Hawkes, H.A. and Jenkins, S.H. (1958) Comparison of four grades of filter media in relation to purification, film accumulation and fauna of sewage percolating filters operating on alternating double filtration. *Journal and Proceedings of the Institute Sewage Purification*, (2), 221–225.

Hawkes, H.A. and Shepherd, M.R.N. (1971) The seasonal accumulation of solids in percolating filters and attempted control at low frequency dosing. *Proceedings of the Fifth International Water Pollution Conference.* Pergamon, Oxford.

Hawkes, H.A. and Shepherd, M.R.N. (1972) The effect of dosing frequency on the seasonal fluctuations and vertical distribution of solids and grazing fauna in sewage percolating filters. *Water Research*, **6**, 721–730.

Hayward, C.H. and Cross, T.S.C. (1976) Design and construction of circular biological filter walls. CIRIA Report No. 58. Construction Industry Research and Information Association, London.

Hemming, M.L. (1979) General biological aspects of wastewater treatment including the deep shaft process. *Water Pollution Control*, **78**, 312–325.

Honda, Y. and Matsumoto, J. (1983) The effect of temperature on the growth of microbial film in a model trickling filter. *Water Research*, **17**, 375–382.

Hoyland, G. and Harwood, N. (1980) Design of biological filtration plants. *Water Pollution Control*, **79**, (3), 352–360.

Hoyland, G. and Roland, D. (1984) Biological filtration of finely-screened sewage. Technical Report TR198. Water Research Centre, Stevenage.

Institute of Water Pollution Control (1972) *Directory of Municipal Wastewater Treatment Plants, Vols. I–IV.* Institute of Water Pollution Control, Maidstone.

Institution of Public Health Engineers (1978) *The Public Health Engineering Data Book 1978–9* (ed. R.E. Bartlett). Sterling Professional Publications, London.

Jackson, C.J. and Lines, G.T. (1972) Measures against water pollution in the fermentation industries. *Pure and Applied Chemistry*, **29**, 381–393.

Kalmykova, Y., Strömvall, A.-M., Rauch, S. and Morrison, G. (2009) Peat filter performance under changing environmental conditions. *Journal of Hazardous Materials*, **166**, 389–393.

Kennedy, P. and Van Geel, P.J. (2010) Hydraulics of peat filters treating septic tank effluent. *Transport in Porous Media*, **41**, 47–60.

Kilcannon, D.F. and Stover, E.L. (1982) Design methodology for fixed-film reactors — RBCs and biological towers. *Civil Engineering for Practicing and Design Engineers*, **2**, 107–116.

Kirjanova, A., Rimeika, M. and Dauknys, R. (2011) Start-up of trickling filters using novel filter medium under low temperature conditions. *Environmental Engineering, the 8th International Conference*, Vilnius Gediminas Technical University, Vilnius, Lithuania, 19–20 May 2011, pp. 578–583.

Koiv, M., Vohla, C., Motlep, R., Liira, M., Kirsimae, K. and Mander, U. (2009) The performance of peat-filled subsurface flow filters treating landfill leachate and municipal wastewater. *Ecological Engineering*, **35**, (2), 204–212.

Laidani, Y., Hanini, S. and Henini, G. (2010) Valorization of *Luffa cylindrica* for water treatment copper chargers: Study of the possibility of regeneration by chemical desorption. *Proceedings of ICAMMM 2010*, Sultan Qaboos Universisty, Oman, December 2010.

Lamb, R. and Owen, S.G.H. (1970) A suggested formulae for the process of biological filtration. *Water Pollution Control*, **69**, (2), 209–220.

Lavender, P.A. (1970) Percolating filter media properties. Report to Imperial College, London, in partial fulfilment of requirements for Diploma of Membership of Imperial College.

Learner, M.A. (1975a) The ecology and distribution of invertebrates which inhabit the percolating filters of sewage-works. Ph.D. Thesis. University of London.

Lens, P.N., Vochten, P.M., Speleers, L. and Verstraete, H. (1994) Direct treatment of domestic wastewater by percolation over peat, bark and woodchips. *Water Research*, **28**, (1), 17–26.

Levine, M., Luebbers, R., Galligan, W.E. and Vaughan, R. (1936) Observations on ceramic filter media and high rates of filtration. *Sewage Works Journal*, **8**, 701–727.

Logan, B.E., Hermanowicz, S.W. and Parker, D.S. (1987) A fundamental model for trickling filter process design. *Journal of the Water Pollution Control Federation*, **59**, (12), 1029–1042.

Lopez-Ponnada, E.V., Lynn, T.J., Peterson, M., Ergas, S.J. and Mihelcic, J.R. (2017) Application of denitrifying wood chip bioreactors for management of residential non-point sources of nitrogen. *Journal of Biological Engineering*, **11**, (16). https://doi.org/10.1186/s13036-017-0057-4

Lu, H., Chandran, K. and Stensel, D. (2014) Microbial ecology of denitrification in biological wastewater treatment. *Water Research*, **64**, 237–254.

Martin, D.G. and Anderson, B.C. (2007) Performance evaluation, comparison and modelling of passive wastewater treatment systems. *Journal of Environmental Engineering and Science*, **6,** (6), 691–702.

McWhinnie, M.J., Chapman, S.J., Edwards, A.C. and Cook, Y.E.M. (2004) A rapid method for the estimation of rock surface areas. *Hydrobiologica*, **517**, (1–3), 133–136.

Mehta, D.S., Davis, H.H., and Kingsbury, R.P. (1972) Oxygen theory in biological treatment plant design. *Journal of Sanitary Engineering Division*, **98**, (SA3), 471–489.

Mills, H.F. (1890) Filtration of sewage and water, and chemical precipitation of sewage at Experiment Stations of the State Board of Health of Massachusetts in 1888 and 1889. Report on water and supply and sewerage: Part II. Massachusetts State Board of Health, Wright and Potter Printing, Co., Boston.

Mondal, B. and Warith, M. A. (2008) Use of shredded tire chips and tire crumbs as packing media in trickling filter systems for landfill leachate treatment. *Environmental Technology*, **29**, 827–836.

Moncrieff, D.S. (1953) The effect of grading and shape on the bulk density of concrete aggregates. *Magazine of Concrete Research*, **5**, 67–70.

Montgomery, H.A.C. and Bourne, B.J. (1966) Modifications to the BOD test. *Journal and Proceedings of the Institute of Sewage Purification*, **65**, 357–369.

Murnane, J. G., Brennan, R. B., Healy, M. G. and Fenton, O. (2016) Assessment of intermittently loaded woodchip and sand filters to treat dairy soiled water. *Water Research*, **103**, 408–415.

National Research Council (1946) Sewage treatment at military installations. *Sewage Works Journal*, **15**, (5), 839–846.

Naz, I. Khatoon, N., Ali, M.I., Saroj, D.P., Ain-ul Batool, S., Ali, N. and Ahmed, S. (2014) Appraisal of the tire derived rubber (TDR) medium for wastewater treatment under aerobic and anaerobic conditions. *Journal of Chemical Technology and Biotechnology*, **89**, (4), 587–596.

Oboh, I.O., Aluyor, E.O. and Audu, T.O.K. (2011) Application of *Luffa cylindrica* in natural form as biosorbent to removal of divalent metals from aqueous solutions — kinetic and equilibrium study. In *Waste Water — Treatment and Reutilization* (ed. F.S.G. Einschlag). InTech, pp. 195–212.

Oliveira, E.A. (2007) Estudo do potencial de utilização da biomassa de Luffa cylindrica na descontaminação de efluentes têxteis contendo íons metálicos e corantes. Ph.D. Thesis. Universidade Estadual de Maringá, Maringá, Brazil.

Parker, D.S. (1999) Trickling filter mythology. *Journal of Environmental Engineering, American Society of Civil Engineers*, **125**, (7), 618–625.

Pearse, P. and Jarvis, S. (2011) Operational experiences with structured plastic media filters: 10 years on. *Water and Environment Journal*, **25**, (2), 200–207.

Pearson, C.R. (1965). The use of synthetic media in biological treatment of industrial wastes. *Journal and Proceedings of the Institute of Sewage Purification*, pp. 519–524.

Pike, E.B. (1978) The design of percolating filters and rotating biological contactors, including details of international practice. Technical Report TR93. Water Research Centre Stevenage, UK.

Porter, K.E. and Smith, E. (1979) Plastic-media biological filters. *Water Pollution Control*, **78**, (3), 371–381.

Raj, S.A. and Murthy, D.V.S. (1998) Nitrification of synthetic wastewater in a cross flow medium trickling filter. *Bioprocess Engineering*, **19**, 149–154.

Raj, S.A. and Murthy, D.V.S. (1999) Comparison of the trickling filer models for the treatment of synthetic dairy wastewater. *Bioprocess Engineering*, **21**, (1), 51–55.

Ramsden, I. (1972) East Keveston RDC sewage treatment programme. *Surveyor*, **140**, 30–31.

Reyes-Lara, S. and Reyes-Mazzoco, R. (2009). Efecto de las cargas hidraulica y org áanicá sobre la eficiencia de remocion de un empaqué estructurado en un filtro percolador (Effect of hydraulic and organic load on the mass removal of a structured packing in a trickling filter). *Revista Mexicana de Ingeniería Química*, **8**, 101–109.

Riahi, K., Mammou, A.B. and Thayer, B.B. (2009) Date-palm fibers media filters as a potential technology for tertiary domestic wastewater treatment. *Journal of Hazardous Materials*, **161**, (2–3), 608–613.

Roberts, J.C. (1973) Towards a better understanding of high-rate biological film flow reactor theory. *Water Research*, 7, 1561–1588.

Rowlands, C.L. (1979) Ecological and comparative performance studies on high-rate filter media. Ph.D. Thesis. University of Aston in Birmingham, UK.

Royal Commission on Sewage Disposal (1908) *5th Report*. HMSO, London.

Ruiz-Marin, A., Campos-Garcia, S., Zavala-Loría, J., Solana, F. and Canedo-López, Y. (2009). Assessment of *Luffa cyllindrica* as support in biofilms reactors for the biological treatment of domestic wastewater. *Water Air Soil Pollution*, **199**, (1), 13–21.

Saroj, S., Singh, S. and Mohan, D. (2015) Removal of colour (direct blue 199) from carpet industry wastewater using different biosorbents (maize cob, citrus peel and rice husk). *Arabian Journal for Science and Engineering*, **40**, (6), 1553–1564.

Schroepfer, G.J. (1951) Effect of particle shape on porosity and surface area of trickling filter medium. *Sewage and Industrial Wastes*, **23**, 1356–1366.

Schultze, K.L. (1957) Experimental vertical screen trickling filter. *Sewage and Industrial Wastes*, **29**, 458–467.

Sharvelle, S. McLamore, E. and Banks, M.K. (2008) Hydrodynamic characteristics in biotrickling filters as affected by packing material and hydraulic loading rate. *Journal of Environmental Engineering*, **134**, (5), 346–352.

Shephard, M.R.N. (1967) Factors influencing the seasonal accumulation of solids in bacteria beds. M.Sc. Thesis. University of Aston in Birmingham, UK.

Shephard, M.R.N. and Hawkes, H.A. (1976) Laboratory studies on the effects of temperature on the accumulation of solids in biological filters. *Water Pollution Control*, **75**, (1), 58–72.

Smith, P.A. and Coackley, P. (1983) A method for determining specific surface area of activated sludge by dye adsorption. *Water Research*, **17**, 595–598.

Solbe, J.F. de L.G., Ripley, P.G. and Tomlinson, T.G. (1974) The effect of temperature on performance of experimental percolating filters with and without mixed macro-invertebrate populations. *Water Research*, **8**, 557–573.

Sousa, J.T., Henrique, I.N., Oliveira, R., Lopes, W.S. and Leite, V.D. (2008) Nitrification in a submerged attached growth bioreactor using *Luffa cylindrica* as solid substrate. *African Journal of Biotechnology*, 7, (15), 2702–2706.

Stanbridge, H.H. (1972) Historical account of rotating and travelling distributors for biological filters. *Water Pollution Control*, **71**, 573–577.

Stanbridge, H.H. (1976) *History of Sewage Treatment in Britain. Biological Filtration, 1.* Institute of Water Pollution Control, Maidstone.

Surakka, S. and Kamppi, A. (1971) Infiltration of wastewater in peat soil. *Suo*, **22**, (3–4), 51–58.

Suschka, J. (1987) Hydraulic performance of percolating biological filters and consideration of oxygen transfer. *Water Research*, **21**, 865.

Tang, X.Y., Zhou, Y.B., Xu, Y.X., Zhao, Q., Zhou, X.A. and Lu, J. (2010) Sorption of polycyclic aromatic hydrocarbons from aqueous solution by hexadecyltrimethylammonium bromide modified fibric peat. *Journal of Chemical Technology and Biotechnology*, **85**, (8), 1084–1091.

Tao, J., Mancl, K.M. and Tuovinen, O.H. (2009) Treatment of sanitary sewer overflow with fixed media bioreactors. *Applied Engineering in Agriculture*, **25**, (1), 39–43.

Tariq, M.N. (1975) Retention time in trickling filters. *Progress in Water Technology*, 7, (2), 225–234.

Thompson, T.J. (1925) Percolating bacteria beds. *Proceedings of the Association of Managers Sewage Disposal Works*, pp. 52–56.

Tomlinson, T.G. (1946b) The growth and distribution of film in percolating filters treating sewage by single and alternating double filtration. *Journal and Proceedings of the Institute of Sewage Purification*, pp. 168–183.

Tomlinson, T.G. and Hall, H. (1950) Some factors in the treatment of sewage in percolating filters. *Journal and Proceedings of the Institute of Sewage Purification*, pp. 338–355.

Tomlinson, T.G. and Hall, H. (1953) Treatment of sewage by alternating double filtration with and without settlement of the primary effluent. *The Water and Sanitary Engineer*, **4**, 86–88.

Tomlinson, T.G. and Snaddon, D.H.M. (1966) Biological oxidation of sewage by films of micro-organisms. *International Journal of Air and Water Pollution*, **10**, 865–881.

Truesdale, G.A., Wilkinston, R. and Jones, K. (1962) A comparison of the behaviour of various media in percolating filters. *Journal and Proceedings of the Institute of Sewage Purification*, (4), 325–340.

Ulug, S. E. and Ucuncu, A. (1992) Process efficiency characterization of plastic medium trickling filter. *International Journal of Environmental Studies*, **42**, 31–40.

Velz, C.J. (1948) A basic law for the performance of biological beds. *Sewage Works Journal*, **20**, (4), 607–617.

Vianna, M.R., de Melo G.C.B. and Neto, M.R.V. (2012) Wastewater treatment in trickling filters using Luffa cylindrical as biofilm support medium. *Journal of Urban and Environmental Engineering*, **6**, (2), 57–66.

Vigueras-Cortés, J.M., Villanueva-Fierro, I., Garzón-Zúñiga, M.A., de Jesús Návar-Cháidez, J., Chaires-Hernández. I. and Hernández-Rodríguez, C. (2013) Performance of a biofilter system with agave fiber filter media for municipal wastewater treatment. *Water Science and Technology*, **68**, (3), 599–607.

Viraraghavan, T. and Rana, S.M. (1991) Treatment of septic tank effluent in a peat filter. *International Journal of Environmental Studies*, **37**, 213–225.

Water Research Centre (1981) STOM: Sewage Treatment Optimization Model. Technical Report TR144. Water Research Centre, Stevenage, UK.

WEF (2011) Biofim reactors. WEF Manual of Practice No. 35. Water Environment Federation Press, Alexandria, Virginia, US.

Wheatley, A.D. and Williams, I.L. (1976) Pilot-scale investigations into the use of random-pack plastics filter media in the complete treatment of sewage. *Water Pollution Control*, **75**, (4), 468–486.

Wilkinson, R. (1958) Media for percolating filters. *Surveyor*, **117**, (3433), 131.

Williamson, K. and McCarty, P.L. (1976) A model of substrate utilization by bacterial films. *Journal of the Water Pollution Control Federation*, **48**, 9–24.

Winkler, M.A. (1981) *Biological Treatment of Wastewater*. Ellis-Horwood, Chichester.

Winkler, M.A. and Thomas, A. (1978) Biological treatment of aqueous wastes. In *Topics in Enzyme and Fermentation Biotechnology* (ed. A. Wiseman). Ellis-Horwood, Chichester, pp. 200–279.

Zhou, L. Bai, X., Li, Y. and Ma, P. (2008) Immobilization of micro-organism on macroporous polyurethane carriers. *Environmental Engineering Science,* **25**, (9), 1235–1241.

Zhu J., Kulick F.M., Li L. and Wang L.K. (2016) Municipal and industrial wastewater treatment using plastic trickling filters for BOD and nutrient removal. In *Natural Resources and Control Processes. Handbook of Environmental Engineering, Vol 17* (eds. L. Wang, M.H. Wang, Y.T.Hung and N. Shammas). Springer, Cham, Switzerland, pp. 285–347.

Zhu, S. and Chen, S. (2002) The impact of temperature on nitrification rate in fixed film biofilters. *Aquacultural Engineering*, **26**, (4), 221–237.

# 4
# The Organisms and Their Ecology

Fixed biofilm reactors differ from the activated sludge process in providing an inert base (i.e., medium) on which bacteria and other microflora form a biofilm. Depending on the type of fixed-film reactor (FFR) the media can provide a wide range of ecological habitats and niches ranging from aquatic to terrestrial. Submerged FFRs have similar ecology to activated sludge dominated by Protozoa, Rotifera and some annelids, while rotating biological contactors (RBCs) (Sec. 5.2) have some higher forms, mainly oligochaetes. Percolating filters are unique in supporting a wide range of air-breathing macroinvertebrates that actively graze on the biofilm, creating a complex food web comprising several extra-trophic levels compared with the activated sludge process (Fig. 4.1). The natural habitat of these microorganisms is the surface of stones and other submerged objects in rivers and streams, especially those receiving large organic and nutrient inputs. The macroinvertebrates found in percolating filters are also associated with similar aquatic habitats such as lakes and ponds, wetlands, and other damp environments.

The basic unit of purification in all FFRs is the biofilm, which converts the soluble and suspended organic matter in the influent wastewater into the bacterial and fungal biomass on which the rest of the community depends. Protozoa graze mainly on the biofilm or free-living bacteria, although a few are predators of other protozoans. The remaining meiofauna, mainly nematodes and rotifers, also feed

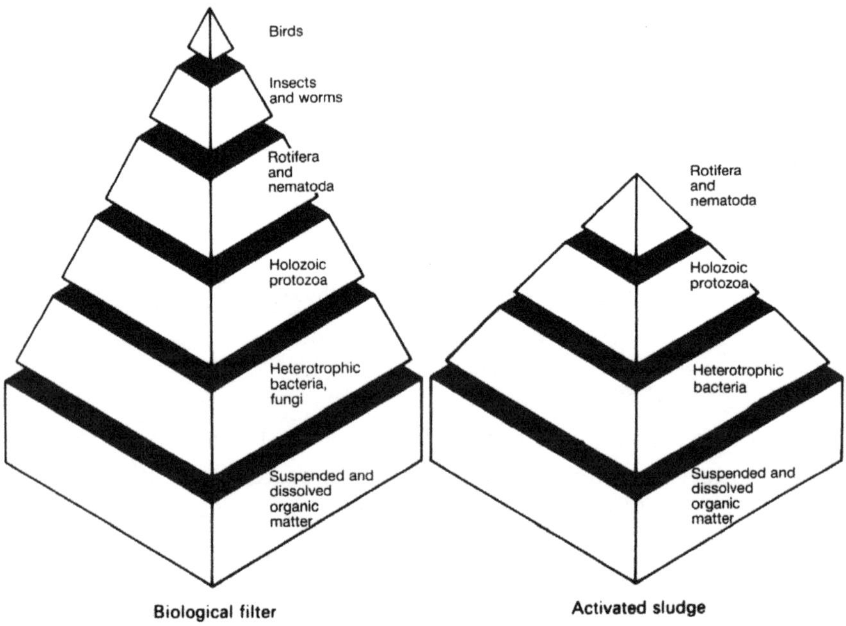


**Fig. 4.1.** Comparison of the food pyramids for a percolating filter and activated sludge treatment systems.

primarily on the biofilm, although several are predatory. The grazing macrofauna, which is dominated by dipteran larvae and oligochaete worms, also feed mainly on the biofilm. The main energy transformations are given in Fig. 4.2. A typical list of species from a percolating filter is given in Table 4.1 and contains species of Bacteria, Fungi, Algae, Protozoa (Sarcomastigophora and Ciliophora), Nematoda, Rotifera, Annelida, Insecta, Crustacea, Arachnida (Acari and Araneida), and Mollusca. This is by no means a definitive list and a vast diversity of species from all phyla has been recorded (Clay, 1964; Learner, 1975a; Curds and Hawkes, 1975). Larger species also regularly visit the surface of the filter beds to feed, especially insectivorous birds such as Wagtails and various small mammals such as rodents and hedgehogs, although access to the mammals is often restricted by the filter bed wall. The ecology of percolating filters has been fully reviewed by Hawkes (1959; 1963; 1983b), with each major group examined in detail by Curds and Hawkes (1975).

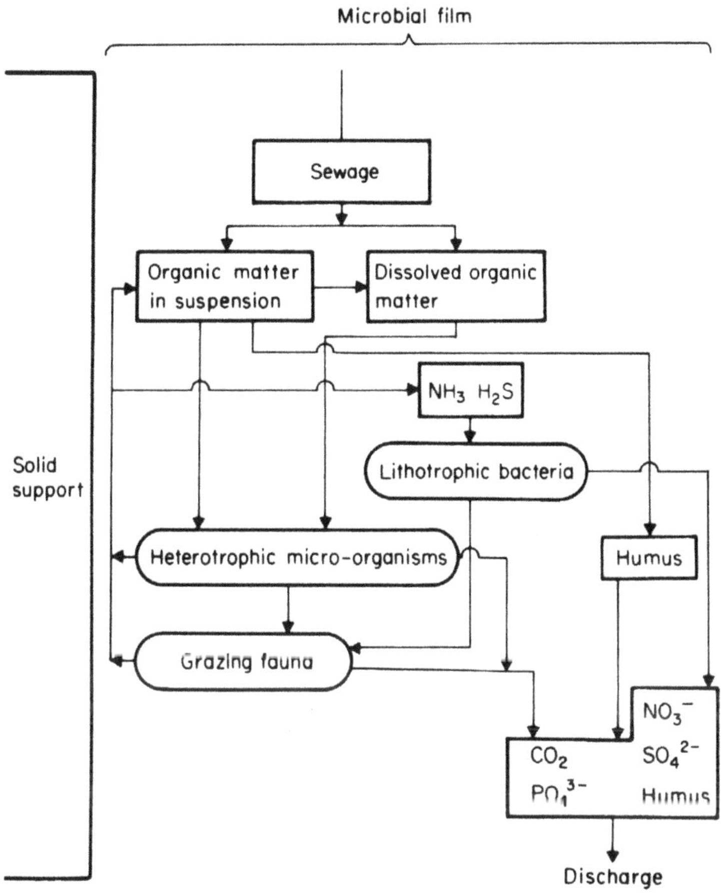


**Fig. 4.2.** Food web in the biofilm of a percolating filter.

**Table 4.1.** Typical list of species recorded using simple microscopy from a single percolating filter.

**Bacteria**

- Zoogloeal forms, mainly *Zoogloeal ramigera*
- *Sphaerotilus* sp.
- *Leptothrix* sp.
- *Beggiatoa* sp.

**Fungi**

- *Subbaromyces splendens* Hesseltine
- *Sepedonium* sp.
- *Fusarium aquaeductuum* (Radimacher and Rabenhorst) Saccardo
- *Geotrichum candidum* Link

(*Continued*)

**Table 4.1.** (*Continued*)

| |
|---|
| **Algae** |
| *Chlorella* sp. |
| *Scenedesmus* sp. |
| *Stigeoclonium* sp. |
| **Protozoa** |
| **Sarcomastigophora** |
| *Bodo* sp. |
| *Amoeba* sp. Mainly *Amoeba radiosa* Ehrenberg |
| *Euglena* sp. |
| **Ciliophora** |
| (Holotrichia) |
| *Trachelophyllum pusillum* Perty-Claparède and Lachmann |
| *Hemiophrys fusidens* Kahl |
| *Hemiophrys pleurosigma* Stokes |
| *Chilodonella cucullulus* (Müller) |
| *Chilodonella uncinata* Ehrenberg |
| *Colpoda cucullus* Müller |
| *Uronema nigricans* (Müller) Florentin |
| *Glaucoma scintillans* Ehrenberg |
| *Colpidium colpoda* Stein |
| *Colpidium campylum* (Stokes) |
| *Paramecium aurelia* Ehrenberg |
| *Paramecium caudatum* Ehrenberg |
| (Peritrichia) |
| *Vorticella microstoma* Ehrenberg |
| *Vorticella convallaria* |
| Linnaeus Vorticella vernalis Stokes |
| *Opercularia minima* Kahl |
| *Opercularia microdiscum* Faure-Fremiet |
| *Opercularia coarctata* Claparède and Lachmann |
| *Epistylis rotans* Svec |
| (Spirotrichia) |
| *Stentor roeseli* Ehrenberg |
| *Aspidisca costata* (Dujardin) = *A. cicada* |
| *Tachysoma pellionella* (Müller-Stein) |
| *Oxytricha ludibunda* Stokes |
| (Suctoria) |
| *Acineta cuspidata* Stokes |
| *Acineta foetida* Maupas |
| *Podophrya maupasi* Bütschli |
| *Podophrya carchesii* Claparède and Lachmann |
| *Podophrya mollis* Bütschli |
| *Sphaerophrya magna* Maupas |

**Table 4.1.** (*Continued*)

| | |
|---|---|
| **Nematoda** | |
| | Unidentified spp. |
| **Rotifera** | |
| (Bdelloidea) | |
| | *Philodina roseola* Ehrenberg |
| (Monogonota) | |
| | Lecane sp. |
| | *Dicranophorus* sp. |
| **Annelida (Oligochaeta)** | |
| Enchytraeidae | |
| | *Enchytraeus buchholzi* Vejdovsky |
| | *Lumbricillus rivalis* Levinsen |
| Lumbricidae | |
| | *Dendrobaena rubida* (Sav.) *f. subrubicunda* (Eisen) |
| | *Eiseniella tetraedra* (Savigny) |
| **Insecta** | |
| (Collembola) | |
| Isotomidae | |
| | *Isotoma olivacea-violacea* gp. |
| (Coleoptera) | |
| Hydrophilidae | |
| | *Cercyon ustulatus* (Prey.) |
| Staphylinidae | |
| | *Unidentified* spp. |
| (Diptera) | |
| Anisopodidae | |
| | *Sylvicola fenestralis* (Scop.) |
| Psychodidae | |
| | *Psychoda alternata* Say. |
| | *Psychoda severini* Tonn. (= *P. albipennis*) |
| Chironomidae | |
| | *Hydrobaenus minimus* Mg. (= *Limnophyes minimus*) |
| | *Hydrobaenus perennis* Mg. (= *Chaetocladius perennis*) |
| | *Metriocnemus hygropetricus* Kieff. (= *M. eurynotus*) |
| Ephydridae | |
| | *Scatella silacae* Lw |
| Sphaeroceridae | |
| | *Leptocera* sp. |

(*Continued*)

**Table 4.1.** (*Continued*)

| |
|---|
| Cordyluridae |
| *Spathiophora hydromyzina* Fall. |
| (Chilopoda) |
| *Lithobius forficatus* Linn. |
| **Crustacea** |
| (Cyclopoida) |
| Cyclopidae |
| *Paracyclops fimbriatus-chiltoni* (Thomson) |
| **Arachnida** |
| (Acari) |
| Acaridae |
| *Histiostoma carpio* (Kramer) |
| Rhizoglyphus echinopus (Fumouze and Robin) |
| Anoelidae |
| *Histiostoma feroniarum* (Dufour) |
| Ascidae |
| *Platyseius italicus* (Bertese) |
| (Araneida) |
| Linyphiidae |
| Unidentified sp. |
| Mollusca |
| (Gastropoda) |
| Limacidae |
| *Agriolimax reticulatus* (Müller) |

## 4.1 Heterotrophs and Phototrophs

Bacteria form the basic trophic level in FFRs and are predominant both in terms of numbers and biomass, with aerobic, anaerobic and facultative species all present. The bacterial flora is similar to that found in activated sludge. These predominately heterotrophic species grow over the inert medium dominating filamentous forms. The dominant aerobic genera are the Gram-negative rods *Zoogloea, Pseudomonas, Achromobacter, Alcaligens,* and *Flavobacterium* (James, 1964; Harkness, 1966; Pike and Carrington, 1972; Pike, 1975) (Sec. 2.3). As in all wastewater bioreactors, it is the heterotrophic bacteria and fungi that are responsible for the primary oxidation of the organic matter in the wastewater. Filamentous bacteria such as *Sphaerotilus natans* and *Beggiatoa* sp. are frequently associated with FFRs, but are rarely dominant. Their abundance is directly associated with organic loading and other specific abiotic factors, although they rarely cause operational problems as they do in activated sludge. However, they can cause localized ponding due to excessive growth. Biofilm development is dependent on the reactor temperature and organic loading, and so is variable during the year in all FFRs. Where the media is completely mixed within the reactor, the microbial composition tends to be largely homogenous as in activated sludge. However, in processes that employ fixed media, then a distinctive variation in microbial composition will be seen with depth.

So, the microbial composition of the biofilm in percolating filters varies seasonally with filter depth and the composition of the wastewater, making it difficult to characterize the microbial community structure. An example of seasonal and depth variation in a pilot filter containing a 50 mm blast-furnace slag medium treating domestic wastewater is given in Fig. 4.3. As the wastewater passes through the filter bed, it is purified by the biofilm, resulting in the wastewater having a different composition at various depths. Consequently, the microbial community structure varies in composition at different depths, having a greater number of species nearer the surface than at the base. This results in a greater diversity of species in percolating filters than in completely mixed systems such as submerged aerated filters and activated sludge. The stratified distribution of species within percolating filters is an additional factor that allows it to withstand shock loadings and be able to deal with wastewaters of varying composition.

Fungi are more common in FFRs and percolating filters, in particular, than in activated sludge, as they are among the first species to colonize the media. Although many fungi can be isolated from FFRs, only a few manage to take advantage of the habitat and flourish (Cooke, 1959). Among those that do flourish are *Geotrichium candidum, Fusarium aquaeductuum, Sepedonium* sp., *Subbaromyces splendens, Ascoidea rubescens*, and *Trichosporon cutaneum* (Fig. 4.4; Table 4.2). Fungi dominate over heterotrophic bacteria under

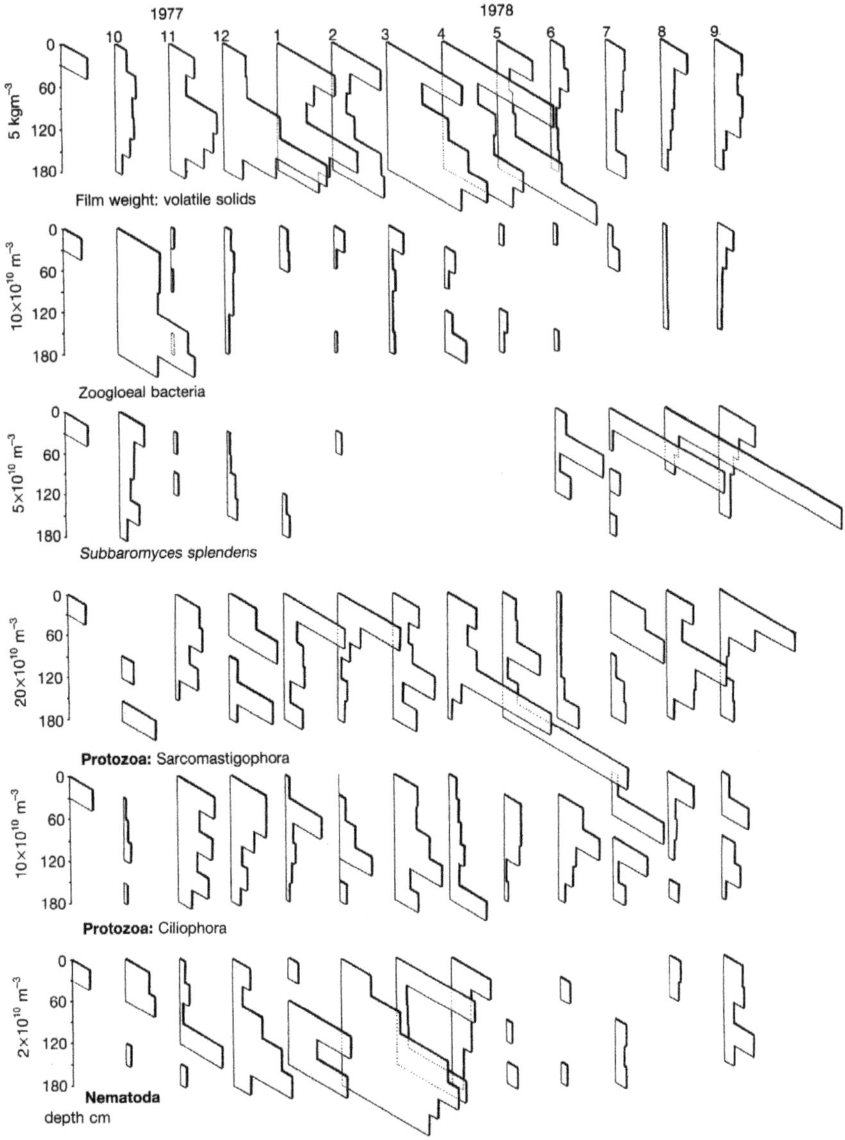


**Fig. 4.3.** Vertical distribution of biofilm and microfauna in a low-rate percolating filter containing 50 mm blast furnace slag.

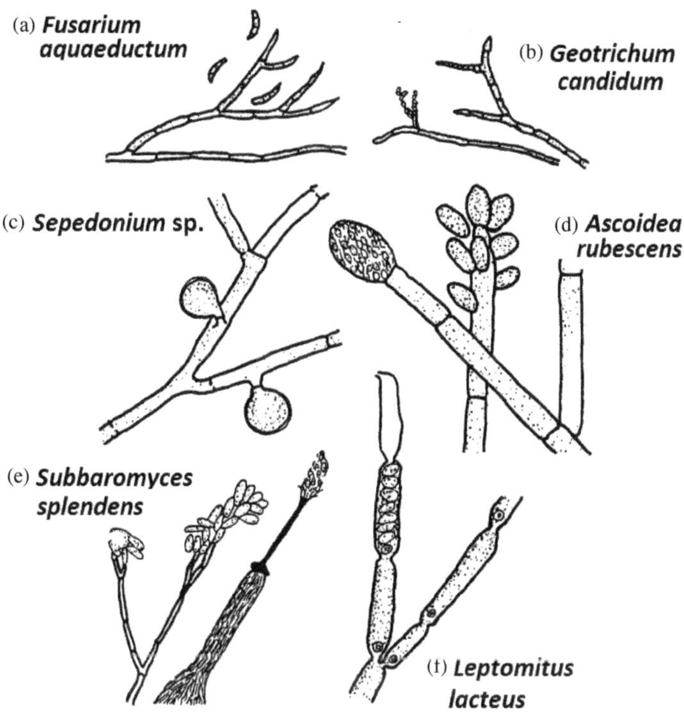


**Fig. 4.4.** The hyphae and spores of some fungi commonly associated with percolating filters (Williams, 1977). Reproduced with permission of the author.

**Table 4.2.** A simple guide to the major fungi occurring in percolating filters.

| |
|---|
| 1. Hyphae |
| Non-septate .... 2 |
| Septate .... 3 |
| 2. Non-septate hyphae with constrictions .... ***Leptomitus lacteus*** |
| 3. Septate hyphae with conidia or arthrospores |
| Absent .... 4 |
| Present .... 5 |
| 4. No conidia or arthrospores formed but chlamydospores present .... ***Sepedonium*** **sp.** |

(*Continued*)

**Table 4.2.** (*Continued*)

5. Conidia formed
   - (i) Conidia boat-shaped and septate (25–35 × 4–5 μm) .... ***Fusarium aquaeductuum***
   - (ii) Conidia formed singly, oval (30–40 × 10–15 5 μm), non-septate, produced sparingly on branched conidiophores .... ***Subbaromyces splendens***
   - (iii) Conidia formed successively at conidiophore apex to form tight clusters, condiophore unbranched, conidia less oval than *S. splendens* but more spherical (20–35 × 12–15 μm), endoconidia formed .... ***Ascoidea rubsescens***

Arthrosphores formed
   - (i) Blastospores present .... ***Trichosporon cutaneum***
   - (ii) Blastospores absent .... ***Geotrichium candidum***

specific conditions. They outgrow bacteria at colder temperatures, under conditions of low pH such as wastewaters from fruit and vegetable processing, or effluents containing mineral acids, and dominate filters treating strong wastewaters with high organic content. For this reason, fungi are associated mainly with industrial effluents (Hesseltine, 1953; Watson *et al.*, 1955; Hawkes, 1957; 1965; Tomlinson, 1946b; Sladká and Ottova, 1968). Fungi have a lower surface area to volume ratio compared to bacteria and thus have a lower affinity for available substrate than heterotrophic bacteria. Therefore, bacteria will predominate when the substrate concentration is low, i.e., when the wastewater is weak or at lower depths in the filter bed after a proportion of the substrate has been utilized and removed. Conversely, fungi dominate when the substrate concentration is high and usually at the surface of the filter bed before the strength of the influent has been reduced. Fungi can attach strongly to media and are not easily dislodged by the shear forces generated by the flow of wastewater. The hyphae protrude from the biofilm into the wastewater flow, and the mycelium physically entrains solids, and fungal-dominated biofilms grow rapidly compared to bacterial biofilms that rely on adsorption only. This results in very thick layers of fungal biofilm developing, which results in the interstices of the medium becoming clogged and ponding often results, due to the reduced grazing activity during winter. Fungal-dominated biofilms are less likely to be nutrient-limited as the nutrients can diffuse throughout the biofilm, resulting in continuous growth. Fungal-dominated biofilms are so thick that if solely bacterial in nature, oxygen would become limiting in the deeper regions. However, the nature of the hyphae allows oxygen transfer deep into the biofilm layer by a process of protoplasmic streaming within the hyphae, preventing anaerobic conditions associated with thick bacterial-dominated biofilms (Tomlinson and Snaddon, 1966). Fungi are generally considered undesirable as dominant members of the biofilm community, as they cause solids accumulation and eventual ponding (Hawkes, 1963). Many authors have also associated heavy fungal biofilms in filters with very large populations of fly larvae, which result in fly nuisance problems. However, fungi have the same removal efficiencies as bacteria, although fungi produce a greater biomass per unit BOD removed, thus resulting in faster biofilm accumulation, and eventually a greater sludge production (Water Pollution Research Laboratory, 1955).

Fungi often dominate the biofilm that develops on plastic media and they easily colonize the smooth surface. One species, *Subbaromyces splendens*, grows particularly well on random plastic medium (Figs. 4.5 and 4.6). The fungus is unusual because it has only ever been isolated from percolating filters with its natural habitat unknown (Hesseltine,

**Fig. 4.5.** Random plastic medium with bacterial biofilm overgrown by colonies of the fungus *Subbaromyces splendens*, eventually leading to ponding.

**Fig. 4.6.** Random plastic medium with thin bacterial film with columns containing sampling baskets. In the foreground to the right is the aluminum access tube for biofilm accumulation assessment using a neutron probe.

1953). It is also unusual because it is associated with weak domestic wastewaters (Hawkes, 1965; Hawkes and Shephard, 1972; Wheatley, 1976; Gray, 1983). Fungi are generally found in greatest abundance on the open surface and within the top 150 mm of the filter bed (Hawkes, 1963). They follow a seasonal pattern of abundance, which slowly increases in autumn and reaches a peak in winter or spring, becoming scarce by mid-summer. A comparative study by Gray (1983) showed that *S. splendens* produced more extensive growth on random plastic mediums than on conventional mineral mediums. The fungus was not restricted to the upper 150 mm but produced extensive vegetative growth throughout the depth of the filter, causing localized ponding both at the surface and inside the reactor. An increase in the hydraulic loading normally leads to a reduction in the fungal growth at the surface of a filter, resulting in a more even distribution of biofilm throughout the bed (Hawkes, 1957). This was also the case in the comparative study when the loading was increased from 1.68 $m^3m^{-3}d^{-1}$ (0.28 kg BOD $m^{-3}d^{-1}$) to 3.37 $m^3m^{-3}d^{-1}$ (0.63 kg BOD $m^{-3}d^{-1}$). However, at the higher loading, the total biomass of the fungus in the filter containing the plastic medium increased ninefold, with ponding occurring in the lower half of the filter. Fungal biofilms are structurally stronger than bacterial ones and so less readily sloughed off (Tomlinson and Williams, 1975). It is this aspect that may help fungi, and in particular, *S. splendens*, to dominate in filters containing the smoother plastic medium. Other factors include: a higher medium surface area resulting in less competition for space with bacteria, the lower diversity of niches within plastic media compared with mineral media for supporting macroinvertebrate grazing fauna, and the large voidage permitting a high ventilation rate resulting in the temperature of the bed being significantly cooler than a mineral filter bed and permitting a large diurnal fluctuation in temperature, which will be similar to the variation in the air temperature, thus inhibiting bacterial growth. The ability of *S. splendens* to cause ponding at depth within the filter has a practical significance, apart from problems in operational control. It has generally been accepted that random plastic filter media can be housed in prefabricated or weaker structures than normally used for mineral media, due to the difference in bulk densities (Table 3.2). However, if ponding was to occur at depth, then the bulk density would increase proportionally to the extra volume of sewage retained. Therefore, under extreme conditions, ponding by the fungus can increase the bulk density of a filter (that is housing a random plastic medium) above the designed structural strength of the reactor itself. This has resulted in the structure becoming damaged, weakened and even collapsing. Three control options have been suggested:

(i) *Recirculation or alternating double filtration*: Hawkes (1965) suggests that fungal growths in percolating filters can be controlled by these modifications. However, the study by Gray (1983) did not show that the fungus was restricted at increased hydraulic loadings or that the growth of the fungus was related to the sewage strength at a particular depth.

(ii) *Periodic dosing*: Continuous dosing has been shown to favor the fungus *S. splendens*, and Hawkes and Shephard (1972) demonstrated that the fungus could be controlled by reducing the frequency of dosing. Although surface growths would be controlled, careful consideration would have to be given to the retention time as this may even out the flow at lower depths in the filter, resulting in enhanced growth of the fungus deeper in the filter bed.

(iii) *Design*: The safest but the most expensive precaution is to ensure that the structure housing the medium can withstand maximum possible bulk density, i.e., total saturation with water (Table 3.2), thus making the reactor completely safe. Such a design, however, would make random plastic filter media less attractive economically, because one of the major advantages of using such media is the saving of capital cost of engineering work (Gray, 1983). This has led to the replacement of random plastic media in large industrial units with vertical strip or string media where the weight of the accumulated film does not affect the walls of the filter or the structural stability of the reactor.

The fungal ecology of percolating filters has been reviewed by Cooke (1954; 1963), Becker and Shaw (1955), and Tomlinson and Williams (1975), but remains poorly understood. Sladká and Ottova (1968) have produced an interesting list of the most common fungi found in biological treatment systems, including percolating filters. Fungi biofilms are commonly employed in submerged FFRs in the treatment of complex industrial wastewaters (Sec. 6).

The surface of percolating filters is the only part of the medium exposed to sunlight that allows photosynthetic algae and bacteria to develop, although they are restricted to the top 50 mm depth of the bed. Phytosynthetic species, and algae, in particular, play a very minor role in the purification process but can interfere with the efficient operation of the filter. Algae may be present as thin or dispersed incrustations of green unicellular algae (e.g., *Chlorella* spp.) and diatoms, or filamentous species can form thick luxuriant surface mats covering the surface of the medium (e.g., *Stigeoclonium* sp.). The felt-like algal mats impair distribution, decrease ventilation, and may even cause ponding. Cyanobacteria (blue-green algae) are classified as bacteria even though they are photosynthetic. In percolating filters, they are present both as unicellular (e.g., *Chlorococcum* sp.) and filamentous forms (e.g., *Oscillatoria* sp., *Phormidium* sp.). The major species causing the dark-green, thick surface mats so often seen in percolating filters is the blue-green alga *Phormidium* sp., which forms a characteristic leathery mucilaginous sheet over the medium that impedes both air and liquid movement through the bed (Benson-Evans and Williams, 1975). The only other FFR where photosynthetic organisms are found is on the discs of RBCs. Most RBC units, however, are covered to prevent fly nuisance and reduce odors, but some larger municipal units are open to the atmosphere and exposed to light permitting algal development (Sec. 5.2). There is currently much interest in developing phototrophic biofilm systems for wastewater treatment, and this is discussed further in Chap. 10.

Mosses and liverworts are also occasionally found on the surface of percolating filters. Moss growth can become so extensive that the entire surface of the bed can become affected. The moss traps solids, forming a thick layer of debris in the interstices of the medium, which interferes with percolation. The most frequently occurring species is *Leptodictyum* (=*Amblystegium*) *riparium*, although Hussey (1982) recorded 12 species of moss in a survey of 64 filters. Factors that encourage moss growth include: a rough, pitted medium, wide spacing of the distribution nozzles providing inter-jet zones, a low organic loading, and the absence of strong industrial or inhibitory wastewaters. The ecology of the filter is altered, with significant changes in the grazing fauna occurring due to the presence of the moss (Table 4.3) (Gray, 1984). Raking may not effectively control the moss and may even help to spread it over the rest of the surface of the bed. Herbicides are effective but can also destabilize the community structure of the filter bed by killing other species. Spraying it with sulfate

**Table 4.3.** Comparison of macroinvertebrate densities (number per liter of medium) in the filter medium with and without surface moss growth. The 95% confidence limits and the level of significant difference (*P*) using the Mann–Whitney test statistic U are also given.

| Macro invertebrate group | Medium | | |
|---|---|---|---|
| | **Moss present $\bar{x}$ CL** | **Moss absent $\bar{x}$ CL** | *P* |
| Lumbricidae | 50 ± 8 | 4 ± 6 | < 0.001 |
| Enchytraeidae | 1040 ± 420 | 356 ± 205 | < 0.05 |
| Psychodidae(larvae) | 1514 ± 370 | 809 ± 317 | < 0.05 |
| Tipulidae (larvae) | 11 ± 3 | 0 ± 0 | < 0.001 |

of iron is also effective but does not permanently remove the moss. Success has been obtained by excluding the light from the surface of the bed or by using a flame gun to burn off the moss. Increasing the organic strength of the influent by reducing the number of filters in operation eliminates moss growth over a period of six months, although there is also normally a deterioration of the final effluent quality. Control can be obtained by increasing the frequency of dosing, reducing the inter jet spacing on the distributors, or by using splash plates to eliminate the inter-jet zones on the surface of the medium. Replacing or covering the surface layer of medium with a smooth surfaced random plastic filter medium will also discourage growth. Prevention is better than trying to eradicate extensive moss growth. Thus, when the surface of the filter is checked periodically for accumulated inorganic debris (plastic strips, etc.), any moss can be weeded out before it becomes established.

## 4.2 Nitrification

Nitrification is a two-stage process with ammonia oxidized to nitrite by bacteria of the genus *Nitrosomonas*, and nitrite to nitrate by *Nitrobacter* spp. (Chap. 7). In low-rate single-pass filters containing a 50 mm stone medium, virtually full nitrification can be obtained throughout the year with a specific ammonia removal rate of between 120–180 mg $m^{-2}d^{-1}$ when loaded with domestic wastewater at 0.1 kg BOD $m^{-3}d^{-1}$, resulting in a final effluent low in ammonia but rich in nitrate. In percolating filters, nitrifying bacteria tend to become established later than heterotrophs, with *Nitrosomonas* becoming established before *Nitrobacter*, as ammonia is abundant. Therefore, the first sign of nitrification in a filter is the production of nitrite rather than nitrate. The number of nitrifying bacteria and the level of nitrifying activity increase with depth (Tomlinson, 1942; Harkness, 1966; Painter, 1970), resulting in the upper level of single-pass filters being dominated by heterotrophs, and the lower section containing a proportionately higher number of nitrifying bacteria. The reason for this apparent stratification is due to several factors. The autotrophic bacteria responsible for nitrification is slow-growing compared to heterotrophs and have an even more reduced growth rate in competitive situations. Therefore, in the upper layers of the bed where there is abundant organic matter, the heterotrophs will dominate. Nitrifiers are extremely sensitive to toxic compounds in the wastewater, especially heavy metals, and so the presence of such compounds will limit the growth of the bacteria until the compounds have been removed from the wastewater by adsorption on to the heterotrophic biofilm as it passes through the filter. Research has shown that the process is also inhibited when the oxygen concentration in the influent wastewater is limited (Heukelekian, 1947; Hawkes, 1963; Tomlinson and Snaddon, 1966). Nitrifying bacteria are strict aerobes and will be inhibited by reduced aerobic conditions caused by high heterotrophic activity or poor ventilation, and as nitrification is a high oxygen-consuming process, adequate supplies of air are required. Although Painter (1970) showed that organic matter did not *directly* inhibit nitrification, he indicated that the nitrifying bacteria needed to be attached to a stable surface, suggesting inhibition may be due to competition for space. When loadings are increased, the depth of heterotrophic activity is extended deeper within the reactor, with the nitrifying bacteria overgrown by the more rapidly growing heterotrophic bacteria. Nitrification is virtually eliminated by hydraulic loadings of domestic wastewater exceeding 2.5 $m^3m^{-3}d^{-1}$. This enhances heterotrophic growth, which can extend throughout the depth of the bed and exclude the nitrifying organisms (Bruce *et al.*, 1970; Joslin *et al.*, 1971; Bruce *et al.*, 1975). Temperature also has a marked influence on nitrification (Painter, 1970), and the large fluctuations in temperature seen in random plastic filters account for the low degree of nitrification that occurs with such media (Gray, 1980). Although the threshold temperature for the process is 10°C, with domestic wastewater rarely falling to below 12°C (normal annual range 12–18°C), a few degrees reduction in the temperature below 10°C is likely to have a disproportionate reduction on nitrification (Bruce *et al.*, 1975). This is seen in comparative studies between mineral and plastic medium filters, where the high voidage of plastic media allows a greater degree of ventilation and large diurnal changes in temperature similar to the air temperature.

Many workers have noticed discrepancies in the total nitrogen balance of their filters, with the decrease in ammonia concentration not corresponding to the increase in nitrate concentration (Bruce *et al.*, 1975; Hemming and Wheatley, 1979; Gray, 1980). This is due to the ammonia being supplied not only in the applied wastewater but also within the filter from the de-animation of organic nitrogen, endogenous respiration, and from cell lysis. Ammonia, on the other hand, is not only removed by nitrification, but also by volatilization of free ammonia and by metabolism into new cellular material. Gray found that at times of low biofilm accumulation when the grazing fauna population was still large, the ammonia concentration in the final effluent was high. He concluded that the observed increase in the ammonia concentration was due to the excretion products of the grazing fauna. The oxygen profile formed by diffusion through the biofilm results in nitrate ions being lost by denitrification as gaseous nitrogen from the anoxic zone near the biomass–medium interface.

## 4.3 Meiofauna

### 4.3.1 *Protozoa*

Protozoa is a major group of unicellular microorganisms that are well represented in all aerobic treatment systems. Many are motile utilizing flagella, cilia or pseudopodia for movement. In FFRs, three classes of protozoa are common. Sarcomastigophorea is the smallest, being <10 μm in diameter. Known as flagellates, due to movement by one or two whip-like flagella, they can be broadly separated into two groups — the Phytomastigophorea contain chloroplasts and can photosynthesize, while the Zoomastigophorea are heterotrophs. The former group is rarely found in large numbers in FFRs, while species of the latter group are common in all FFR systems. Flagellates are unable to feed on particles but are osmotrophic, i.e., they absorb dissolved nutrients directly through the cell wall (Fig. 4.7). The Sarcodina (amoebae) is a variable group of protozoa with species varying in size from 10 to 250 μm. Amoebae are transparent cells that crawl or stream extremely slowly over surfaces, making them almost

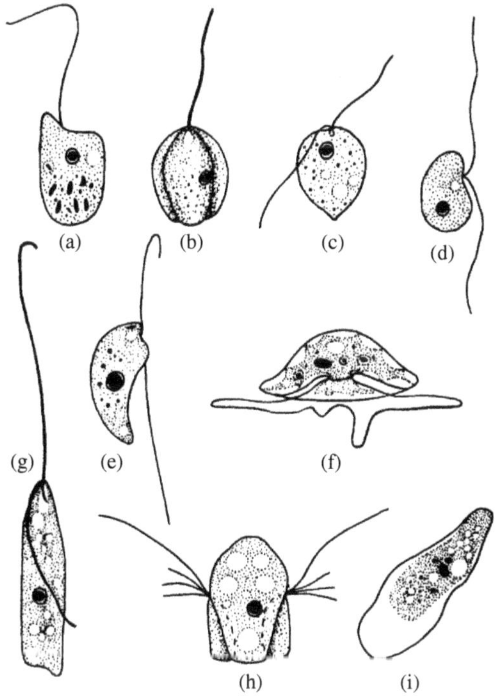


**Fig. 4.7.** Common flagellates and amoeba associated with fixed-film reactors: (a) *Oicomonas termo*, (b) *Notosolenus orbicularis*, (c) *Cercobodo crassicauda*, (d) *Pleuromonas jaculans*, (e) *Bodo caudatus*, (f) *Arcella vulgaris*, (g) *Peranema trichophorum*, (h) *Trepomonas agilis*, and (i) *Vahlkampfia limax* (Curds, 1975). Reproduced with permission of Elsevier Publishing.

indistinguishable from the surrounding debris. Movement is achieved by extending pseudopodia into which the remainder of the cell flows. They feed on bacteria by ingesting them inside contractile vacuoles. Amoebae are either naked or testate (i.e., have a small shell that partially covers or houses the body) (Fig. 4.7). The final group of protozoans, the Ciliatea (ciliates), is the best known due to their relatively large size (20–200 μm). They all process cilia, which are small hairs usually dispersed over the entire surface of their body, although these hairs have become modified to form cirri (tufts) or even tentacles. They predominately feed on bacteria, although some are predatory on other protozoa, and are either free-swimming, attached to the

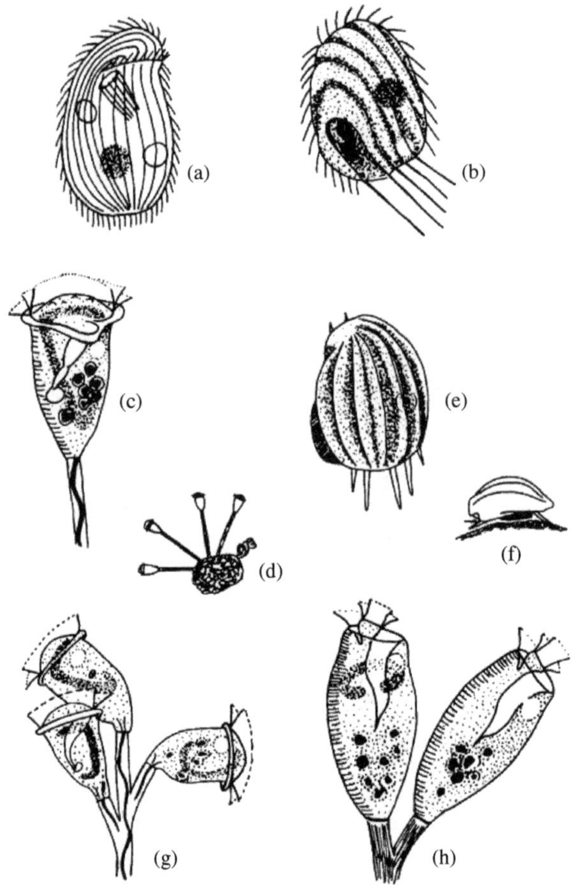


**Fig. 4.8.** Some common ciliates in fixed-film reactors: (a) *Chilodonella unicata*, (b) *Cinetochilum margaritaceum*, (c) *Vorticella convallaria*, (d) *V. convallaria* growing on a sludge floc, (e) *Aspidisca costata*, (f) *A. costata*'s lateral view showing its crawling habit, (g) *Carchesium polypinum*, and (h) *Opercularia microdiscum* (Curds, 1975). Reproduced with permission of Elsevier Publishing.

surface of the biofilm by a stalk, or crawl over the surface of the biofilm. The cilia are used both for locomotion and feeding (Fig. 4.8).

The class ciliatea is divided into four subclasses:

(i) *Holotrichia* are free-swimming protozoans, having bodies that are covered in uniform ciliature. Typical Holotrichia include the genera *Chilodonella*, *Colpoda*, *Colpidium*, *Paramecium* and *Uronema*.

(ii) *Spirotrichia*, in contrast, do not have cilia uniformly distributed over their bodies, but the cilia are bound together to form thick tufts known as cirri. Among the common genera are *Stentor*, *Aspidisca*, *Euplotes*, *Stylonychia* and *Oxytrichia*.

(iii) *Peritrichia* are sessile species with no cilia but conspicuous oral ciliatures on their bodies. Each body is barrel or inverted bell-shaped and is borne on a stalk that may be contractile. In some species, the body can break free from the stalk to form a free-swimming form (a telotroch). By closing the mouth, the telotroch can use the ring of aboral cilia to propel the body through the water. Typical genera are *Vorticella*, *Carchesium*, *Opercularia* and *Epistylis*.

(iv) *Suctoria* are predatory and do not possess cilia or locomotary organs; they catch other protozoans with tentacles. Two types of tentacles are seen: sharp tentacles that capture prey by piercing its body, or tentacles that hold the prey by suction via an adapted terminal knob. Typical genera include *Tokophyra*, *Acineta*, *Podophyra* and *Sphaerophrya*.

The taxonomy and identification of the protozoans are dealt with by Curds (1969). Examples of typical ciliates found in percolating filters are shown in Fig. 4.8.

Protozoa are particularly abundant in percolating filters, and over 218 species have been isolated. There are 35 species of Phytomastigophorea, 30 species of Zoomastigophorea, 31 species of Rhizopodea and 7 species of Actinopodea. However, the bulk (~116) of the species belong to the class Ciliatea. The abundance of ciliates, amoebae and flagellated species in filters has resulted in most studies concentrating on them.

The role of protozoans in all FFRs is similar to that of the activated sludge process, with species feeding mainly on free-living bacteria and clarifying the effluent as well as stimulating bacterial growth by reducing the population density. Some protozoans can feed on non-living particulate, or soluble organic material, whereas a few species are predators of other protozoans. It is generally thought that the ciliates are numerically dominant over the flagellates (Brink, 1967), although Barker (1942; 1946) and Gray (1980) both found flagellates to be dominant in percolating filters. Gray found the mean ratio of Sarcomastigophora to Ciliophora to range from 3.7 to 4.3 in low- and high-rate mineral media filters and 2.0 to 4.4 in low- and high-rate random plastic media filters, respectively. However, flagellates are significantly smaller than ciliates both in overall dimensions and in volume, therefore, in terms of biomass, ciliates are always dominant. Population densities of protozoa increase with organic loading, although the flagellate and free-swimming ciliate protozoans are more susceptible to being flushed out of the plastic medium filter at higher loading. It has been reported that the number of ciliate species in filters increases with depth (Baker, 1949; Baines *et al.*, 1953; Curds, 1975; Wheatley, 1976), although conversely, Hussey (1975) observed that ciliate diversity was negatively associated with the biofilm weight. Many of these workers suggested that ciliates are unable to compete effectively with the other organisms normally associated with the biofilm in the low-dissolved oxygen conditions prevalent at the filter surface; reporting the greatest diversity and largest abundance of ciliates in the lower depths of filters where there is an increasingly smaller concentration of organic matter in the partially treated wastewater. Vertical distribution of various ciliate species within filter beds has been reported (Lackey, 1924; 1925; Frye and Becker, 1929; Cutler *et al.*, 1932; Holtje, 1943; Barker, 1946; Ingram and Edwards, 1960). Generally, they found that particular species tended to predominate at certain depths, with stratification being dependent on factors such as nutrition, or a mixture of environmental and biological interactions. Liebmann (1951) showed that whereas bacterial feeding species were present throughout the bed, being especially abundant in the upper areas, predatory species of protozoans were restricted to the lower half of the filter. Liebmann (1949) also put forward the saprobity theory that as the wastewater passed through the filter it was gradually purified so that the saprobic nature of the liquid environment changed with depth. This was polysaprobic at the top, followed by α- and β-mesosaprobic zones as the depth increased. Thus, for example, *Vorticella microstoma*, *Glaucoma scintillans* and *Colpidium colpoda* are restricted to the surface of the filter (polysaprobic), whereas *Paramecium caudatum*, *Chilodonella uncinata*, *Uronema nigricans*, *Opercularia coarctata* and *Podophyra fixa* are all typical of the middle regions of the filter (α-mesosaprobic). *Aspidisca costata* is usually recorded in the lower regions of filters and is associated with the lowest levels of organic matter (β-mesosaprobic). Gray (1980) found that at the lower loading, ciliates were widely distributed throughout the depth of his pilot filters, and although certain species were limited to specific regions, there was generally no increase in either diversity of species or abundance with depth. At the higher loading, maximum species diversity and the abundance of protozoans occurred in the lower half of the filters. The increased abundance of certain species in the lower depths of the filters at this loading, such as *Opercularia microdiscum*, was not due to the lower organic content of the influent sewage at that depth as suggested previously, but because this species is tolerant of both organic load and biofilm accumulation. It is more probable that the increased occurrence of protozoans at the lower depth was a consequence of the increased hydraulic loading. This is confirmed by a greater number of individuals being washed out with the final effluent at the higher loading. In conventional filters, there would be a greater tendency for the protozoan fauna to be forced deeper into the filters by using the normal instantaneous and heavy system of sewage application, compared to the finer distribution system used by Gray and Learner (1983) on their field-scale pilot filters.

In a survey of 52 percolating filters in the United Kingdom (UK), Curds and Cockburn (1970) found that all the filters including those treating industrial wastewater contained protozoans, with ciliates generally dominant. They found that the protozoan fauna of percolating filters closely resembled those of activated sludge but that certain species were found in either one or the other of the processes (Table 4.4). The ciliates identified in the survey consisted of 19 holotrichs, 20 peritrichs, 11 spirotrichs and 3 suctorians. The most common ciliates found are listed in Table 4.5, with *C. uncinata* present in

**Table 4.4.** Comparison of characteristic ciliate protozoans identified in percolating filters and activated sludge plants (Curds and Cockburn, 1970). Reproduced with permission of Elsevier Publishing.

| Percolating filters | Activated sludge plants |
|---|---|
| *Opercularia microdiscum* | *Aspidisca costata* |
| *Carchesium polypinum* | *Vorticella convallaria* |
| *Vorticella convallaria* | *Vorticella microstoma* |
| *Chilodonella uncinata* | *Trachelophyllum pusillum* |
| *Opercularia coarctata* | *Opercularia coarctata* |
| *Opercularia phryganeae* | *Vorticella alba* |
| *Vorlicella striata* var. *octava* | *Carchesium polypinum* |
| *Aspidisca costata* | *Euplotes moebiusi* |
| *Cinetochilum margaritaceum* | *Vorticella fromenteli* |

**Table 4.5.** Most frequently observed species of protozoans recorded in percolating filters (Curds and Cockburn, 1970). Reproduced with permission of Elsevier Publishing.

| Class | Species |
|---|---|
| Phytomastigophorea | *Peranema trichophorum* |
| Zoomastigophorea | *Bodo caudatus* |
| | *Trepomonas agilis* |
| Rhizopodea | Small amoebae |
| | *Arcella vulgaris* |
| Ciliatea | *Aspidisca costata* |
| | *Carchesium polypinum* |
| | *Chilodonella uncinata* |
| | *Cinetochilum margaritaceum* |
| | *Opercularia coarctata* |
| | *Opercularia microdiscum* |
| | *Trachelophyllum pusillum* |
| | *Vorticella convallaria* |
| | *Vorticella striata* var. *octava* |

90% of the samples, *Vorticella convallaria* in 83%, *O. microdiscum* in 81%, and *Carchesium polypinum* in 62%. However, although species are widely distributed, they are not always abundant or the dominant protozoan in a filter. Curds and Cockburn only recorded eight ciliates in large numbers: *C. uncinata* (4% of sites), *C. polypinum* (15%), *Vorticella alba* (2%), *V. convallaria* (10%), *Vorticella striata* var. *octava* (2%), *O. coarctata* (2%), *O. microdiscum* (44%), and *Opercularia phryganeae* (4%). Species were restricted by a variety of factors, including saprobity, flow rate, biofilm accumulation, season, and medium type.

Gray (1980) found that five species (four holotrichs, *Paramecium aurelia*, *Uronema nigricans*, *Glaucoma scintillans*, and *C. colpoda*, and one peritrich *O. microdiscum*) dominated his filters and that all feed on bacteria (Table 4.6). Under low-rate conditions, *O. microdiscum, U. nigricans* and *P. aurelia* were most frequently recorded, whereas, for those under high-rate conditions, only *O. microdiscum* was regularly recorded as abundant (Fig. 4.9). *Paramecium aurelia* is inhibited by high hydraulic loadings, most likely due to its relatively large size (120–150 μm in length). The abundance of this species is linked to biofilm accumulation, being more abundant in the winter months and scarce when the biofilm is at its minimum thickness from May to August (Fig. 4.10). At the low-rate loading, *P. aurelia* was recorded throughout the bed, where it comprised 70–80% of the total ciliate population at times, whereas, at the higher rate, it was restricted to the lower half of the filter until it was eventually washed out.

**Table 4.6.** Number of months when specific ciliate species were dominant in experimental filters loaded at 1.68 $m^3m^{-3}d^{-1}$ (low-loading) and 3.37 $m^3m^{-3}d^{-1}$ (high-loading) (Gray, 1980).

| | Low loading | | | High loading | | | Total | | |
|---|---|---|---|---|---|---|---|---|---|
| Species | Slag | Filter mixed | Plastic | Slag | Filter mixed | Plastic | Low loading | High loading | Total dominance |
| *Uronema nigricans* | 2 | 5 | 5 | 1 | 1 | 2 | 12 | 4 | 16 |
| *Paramecium aurelia* | 5 | 1 | 4 | 1 | 0 | 1 | 10 | 2 | 12 |
| *Opercularia microdisum* | 5 | 5 | 2 | 8 | 7 | 5 | 12 | 20 | 32 |
| *Colpidium colpoda* | 0 | 0 | 1 | 1 | 2 | 3 | 1 | 6 | 7 |
| *Glaucoma scintillans* | 0 | 1 | 0 | 0 | 1 | 0 | 1 | 1 | 2 |

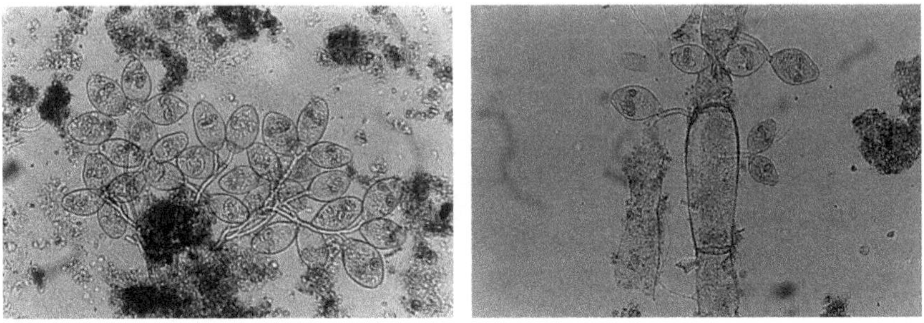

(a) (b)

**Fig. 4.9.** (a) The ciliate protozoan *Opercularia microdiscum* and (b) colonies of *O. microdiscum* attached to the hyphae of the fungus *Subbaromyces splendens.*

*Uronema nigricans* is also unable to cope with high hydraulic loadings but is found in greatest abundance in the summer, with numbers declining as the biofilm accumulates. The density of *Uronema* is negatively correlated with biofilm accumulation, which is in contrast to *P. aurelia*. *Uronema*'s preference for light biofilm development results in *P. aurelia* and *U. nigricans* being rarely found together (Fig. 4.10). *Opercularia microdiscum* is a sessile organism attached to the medium or other suitable substrate by a non-contractile stalk and feeds passively on free-swimming bacteria. It is unable to actively search for food and, unlike the holotrichs, it is unable to move away from any adverse environmental changes, predators or the activities of the macroinvertebrate grazing fauna, except in its telotroch phase. The population density of the opercularian remained relatively small under low-rate conditions, but when the loading was increased to high-rate, the mean population density increased five-fold, occasionally making up between 90–100% of the total ciliate population recorded in all three pilot filters. However, *O. microdiscum* was unable to compete successfully with *U. nigricans* at either loading during periods of light biofilm accumulation but was found in greatest abundance during January/February and July/August, the months when biofilm accumulation was heavy. The sedentary ciliates attach themselves to a variety of substrates including zoogloeal bacteria, fungal hyphae, insect debris and the larger filamentous bacteria. *Opercularia*

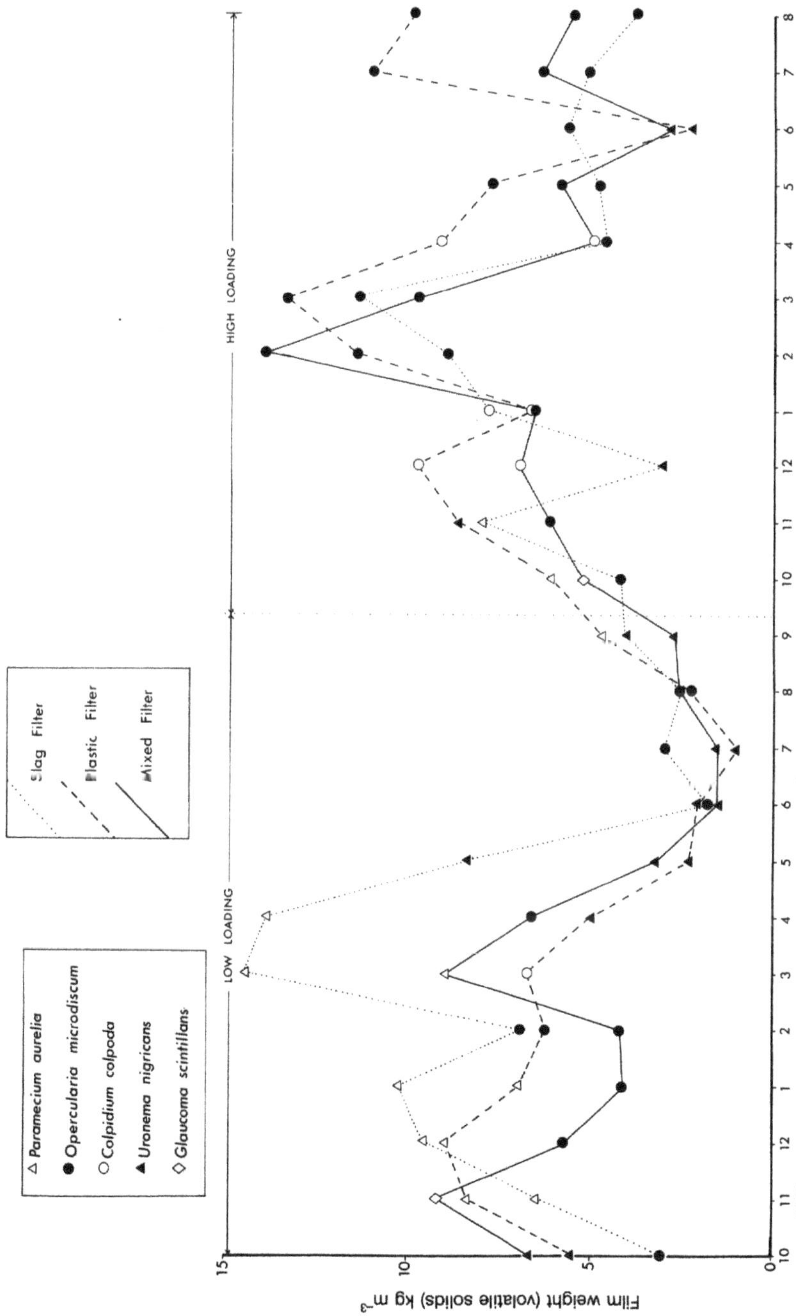


**Fig. 4.10.** Dominant ciliate protozoan species in pilot filters at various biofilm accumulations (Gray, 1980).

*microdiscum* was positively correlated with both zoogloeal bacteria and the fungus *S. splendens*, which it used as a substrate for attachment. In the pilot filters, the peritrich was restricted by competition for food and space under low-rate conditions and limited only by the lack of suitable surfaces for attachment at the higher loading. In all the pilot filters, *O. microdiscum* was found throughout the depth of the beds during the low loading, except at depths with heavy biofilm accumulation or high abundance of *P. aurelia*. With the increase in loading rate, however, the population increased mainly in the lower half of the beds, although still avoiding those areas of heaviest biofilm accumulation. The clear association of the ciliate with *S. splendens*, the tough hyphae of which grow out into the liquid layer flowing over the biofilm, provides an ideal niche for the sessile protozoan so that it is away from any danger of being overgrown by the rapidly developing biofilm (Fig. 4.9). Although *O. microdiscum* has been previously associated with low concentrations of organic matter (Barritt, 1940; Barker, 1946; Tomlinson and Snaddon, 1966), more recent work has suggested that the species may have a more general distribution (Curds, 1969; Curds and Cockburn, 1970; Hussey, 1975). Learner (1975a) also noted the importance of this species in wastewater treatment and found it to be the dominant organism in the majority of filters examined. He recorded a positive correlation of the organic and hydraulic loading with *O. microdiscum*, noting that maximum abundance was recorded in filters receiving loads exceeding 0.25 kg BOD $m^{-3}d^{-1}$. Some years earlier, Bruce and Merkens (1970) had found large numbers of the species, but no other ciliate, in experimental filters receiving an organic loading of 2.0 kg BOD $m^{-3}d^{-1}$ at a hydraulic loading of 6 $m^3m^{-3}d^{-1}$. *Colpidium colpoda* occupies the same niche as *P. aurelia* and is present at the same time. The species is associated with the lower range of biofilm accumulations at which *P. aurelia* dominates, and is found reaching maximum numbers at biofilm weights of between 4 and 9 kg $m^{-3}$. At the higher loading, the density of *C. colpoda* increased due to the exclusion of *P. aurelia*, where it dominated at periods of moderate biofilm accumulation. At the higher loading, the sequence of species from heavy to light accumulation can be seen quite clearly, from *O. microdiscum* to *C. colpoda* and then to *U. nigricans* at the lighter weights of biofilm, with each peak

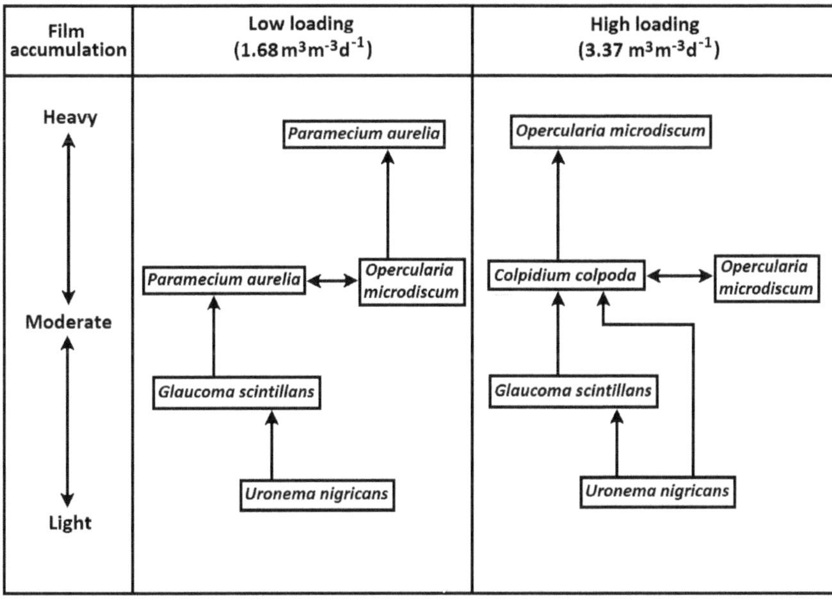


**Fig. 4.11.** Succession of dominant protozoan species with film accumulation (Fig. 4.10) at loadings of 1.68 and 3.37 $m^3m^{-3}d^{-1}$ (Gray, 1980).

in the population density separated from the next. *Colpidum colpoda* and *O. microdiscum* are both high-rate species and compete directly; when one species is found in large numbers, the other is present only in small numbers. It does seem, however, that *O. microdiscum* is more successful than *C. colpoda* when there are enough suitable surfaces for attachment. The succession of these species, which are either controlled by biofilm accumulation or temperature, at low- and high-rate loadings, is summarized in Fig. 4.11. The final dominant species, *Glaucoma scintillans*, competes directly with *U. nigricans* and the former is more successful at moderate biofilm weights (5–9 kg $m^{-3}$), whereas *U. nigricans* is dominant at low biofilm weights (<5 kg $m^{-3}$).

Curds (1973; 1975) divided the wastewater protozoan fauna into three groups according to their habitat: those that swim freely in the liquid phase and are prone to be washed out, those that crawl over the surface of the biofilm and are occasionally washed out, and those that are attached directly to the biofilm, or some other material, and are only removed during sloughing. The habitat preference of those species found in percolating filters is important in survival terms and dictates which species are to be successful. Bungay and Bungay (1968) found that a peritrich, such as *O. microdiscum*, is always present even after sloughing in quite large numbers, and, therefore, potentially able to build up the population rapidly. However, in any given situation, in the competition between species for food, the organism that is fastest both to grow and reproduce under the prevailing conditions will become dominant (Moser, 1958).

*Opercularia microdiscum* and *P. aurelia* were found at all depths by Gray (1980) in his low-rate filters, whereas *C. colpoda* was found in the upper regions of the bed where the organic concentration of the wastewater was greatest. *Aspidisca costata* was limited to the lower half of the pilot filters and was continuously washed out of the filters in large numbers. The dominant species at the higher loading were restricted to particular depths in line with the saprobic index proposed by Liebmann (1949). *Colpidium colpoda* was found in the top and middle regions of the pilot filters, *O. microdiscum* in the middle and lower regions, whereas *A. costata* and *U. nigricans* were restricted to the lower portion of the filters. Suctoria, which are mainly predators on other protozoan species, were found in the middle and lower areas of the filters only where they would have maximum opportunity to come into contact with suitable prey. However, Gray concludes that no individual reason can account for the stratification of the various protozoan species, but that it is the result of a number of environmental (e.g., nature of wastewater, organic load, temperature, hydraulic flow, food availability, surface area) and biological factors (e.g., competition, predation, type of biofilm) that change continuously and alter the distribution of the protozoans within the biofilm. Mistri *et al.* (1994) have further studied the effects of physico-chemical and plant operational variables on a protozoan's community structure.

There are a number of excellent keys for the identification of protozoa in FFRs (Curds, 1969; Foissner and Berger, 1996).

### 4.3.2 *Rotifera*

The role of rotifers in wastewater treatment processes has been reviewed by Doohan (1975). Unfortunately, most of the research carried out on this group has been in connection with the activated sludge process (Curds and Vandyke, 1966; Calaway, 1968; Sydenham, 1968; 1971), thus, relatively little is known concerning the rotifers found in the percolating filter environment (Donner, 1966). In the completely mixed environment of the activated sludge process, rotifers have two distinct functions — they break up floc particles providing nuclei for new floc formation and they clarify the effluent by removing non-flocculated bacteria that are in suspension. They are powerful feeders, and the strong ciliary currents produced by rotifers allow them to feed effectively, even when the concentration of bacteria on which they feed is very low. This is in contrast to protozoans such as Vorticellids, which have weaker ciliary action and therefore are unable to survive at such low bacterial concentrations. Thus, in the activated sludge process, Vorticellids are normally succeeded by rotifers, and for the same reason, rotifers are generally found in the lower sections of percolating filters. From field experiments, Doohan (1975) supposed that it was food availability that exerted the greatest influence on the reproduction rate of the Rotifera, rather than other environmental factors, such as temperature. Therefore, their distribution within the filter environment would appear related to food availability and the ability to compete with other bacterial feeders. Rotifers have greater mobility compared to protozoans, for example, *Lecane* spp. have a specialized foot that enables them to crawl over the biofilm and medium. This specialization allows rotifers to move within the filter bed and extends their distribution area when conditions permit. When protozoans are under stress because of food limitation, rotifers usually move into that area and compete for the available bacteria. Rotifers also help in the production of discrete fecal pellets that consist of undigested material bound together by mucus that rapidly settles. Rotifers of the genus *Lecane* have been shown to feed directly on filamentous bacteria in activated sludge, offering a potential control mechanism of bulking by such bacteria. This includes Types 021N (Kocerba-Soroka *et al.*, 2013); 0092 and *Microthrix parvicella* (Pajdak-Stós *et al.*, 2017); and *Thothrix* spp. and *Halicomenobacter* spp. (Kowalska *et al.*, 2014; 2016). Filamentous bacterial growths are less common in FFRs, but nevertheless, *Lecane* spp., which are common in all FFRs, would also be feeding on filamentous bacteria and

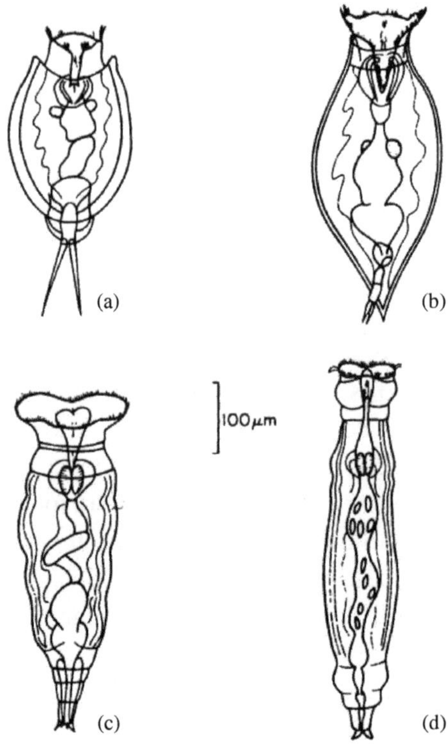


**Fig. 4.12.** Commonly reported rotifers in both fixed-film reactors and activated sludge: (a) *Lecane* sp., (b) *Notommata* sp., (c) *Philodina* sp., and (d) *Habrotrocha* sp. (Doohan, 1975). Reproduced with permission of Elsevier Publishing.

controlling excessive development. This is especially important in integrated fixed-film activated sludge (IFAS) systems or where fixed FFRs are used as a pre-treatment step before activated sludge (Sec. 6.6). Doohan (1975) has provided a comprehensive list of Rotifera species found in wastewater treatment processes, while Pontin (1978) has produced a useful identification key (Fig. 4.12).

### 4.3.3 *Nematoda*

Little is known regarding the role of Nematoda in wastewater treatment processes, but they are present in large numbers in all FFRs, especially percolating filters (Peters, 1930; Lloyd 1945; Calaway,

1968). Of the numerous species recorded in FFRs, *Diplogaster strictus* and *Rhabdites* spp. are amongst the most common (Cooke, 1959; Petersen, 1982). The population of nematodes is dominated by bacterial feeders, although predator species feeding on other nematodes and rotifers are also present, but are far less abundant (Schiemer, 1975). Weninger (1964) recorded maximum population densities of 180 individuals per milliliter, whereas Scherb (1968) found maximum densities of up to 280 $mL^{-1}$ in a bench-scale-activated sludge plant. Schiemer (1975) suggested that experiments carried out by Pillai and Taylor (1968) showed that the maximum population in conventional low-rate filters could be in the order of 1,000 $mL^{-1}$, which corresponds closely with the results obtained by Gray (1980), where a maximum population density of 940 $mL^{-1}$ was recorded in his low-rate pilot filter containing slag medium (Fig. 4.4). Bergtold *et al.* (2007) reported 2,380 and 4,411 individuals' $mL^{-1}$ in the two laboratory biofilters, equivalent to a mean biomass ranging from 209 to 330 mg wet weight per mL. Fast-growing species (e.g., *Paroigolaimella bernensis* and *Diplogasteritus nudicapitatus*) were observed under unstable conditions while the slower-growing species (e.g., *Diploscapter coronatus* and *Acrostichus* sp.) were associated with more stable environmental conditions.

The seasonal variation in abundance coincides with biofilm accumulation, with maximum population densities of nematodes recorded during early spring and with minimum population densities occurring immediately after sloughing. Weninger (1971) and Murad and Bazer (1970) observed that population density was inversely related to temperature, resulting in maximum populations of between 7 and 10°C, although Chaudhuri *et al.* (1965) recorded maximum nematode densities of between 17 and 18°C. Hawkes and Shephard (1972) regarded the Nematoda as being important grazers that were closely associated with biofilm accumulation. Woombs and Laybourn-Parry (1986) studied energy flows in low-rate percolating filters and found that while bacterivore nematodes represented between 0.03 and 2.7% of the total community respiration, their main effect was their grazing impact on bacteria, including pathogens. Nematodes live in the biofilm rather than on the surface or in the liquid phase and thus are directly related to biofilm accumulation and biofilm thickness. They were appreciably less abundant on smoother plastic media where biofilm thickness is less, as compared with rough mineral media. Another advantage of nematodes is their high reproductive potential (Schiemer, 1975), which is similar to that of the astigmatid mites and the enchytraeid worms, with all three groups being able to respond rapidly to increases in the available food. Schiemer considered the role of nematodes in the filter environment and estimated that the nematode population is responsible for 0.001% of the community's respiration. He identified that they had three functions: grazing of bacteria (which affects bacterial density and growth), decomposition of organic matter, and recycling of energy-rich substances inside the filter because of excretion products, feces and dead body tissues, although these may be lost to the system via the final effluent. Thus, although nematodes may have important effects on bacterial activity, they have probably only a minor role in the decomposition of organic matter in wastewater. Identification of nematodes is difficult but there are a number of useful online keys that help separate the key groups, such as the one produced by the Nematology Research Group at the University of Nebraska (https://nematode.unl.edu/key/nemakey.htm).

## 4.4 Grazing Fauna

Percolating filters are dominated by two phyla of grazers, the Annelida and Insecta. Other grazers from the Arachnida, Crustacea and Mollusca are also present but are generally of less significance (Fig. 4.13). Learner (1975a) carried out a detailed survey of invertebrates at 67 separate filter beds at 48 sewage treatment plants throughout the UK. All were sampled in spring, while 59 were also sampled a second time during autumn. A 2.5-liter sample of filter medium and biofilm was removed from the top 300 mm of each filter bed, and a 10-$cm^3$ solids sample was collected from the final effluent using a 400-μm mesh-drift net. A total of 87 species were recorded, of which 39 occurred in the surface samples. From the surface samples, 24 species were particularly successful, being both widespread and abundant (Table 4.7). Three species, the worm *Lumbricillus rivalis*, the larvae of the moth fly *Psychoda alternata*, and the mite

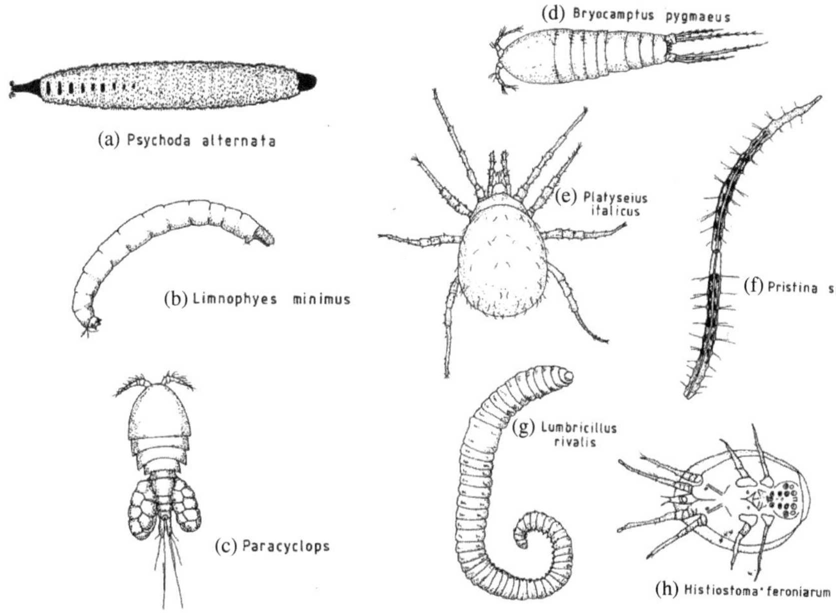


**Fig. 4.13.** Common invertebrates in percolating filters with approximate lengths in parenthesis: (a) *Psychoda alternata* larva, dorsal view (7 mm), (b) *Limnophyes minimus* larva, lateral (3.5 mm), (c) *Paracyclops* sp., dorsal view of female with egg sacs (0.8 mm), (d) *Bryocamptus pygmaeus* female, dorsal (0.5 mm), (e) *Platyseius italicus* female, dorsal (0.8 mm), (f) *Pristina* sp. showing asexual budding (5 mm), (g) *Lumbricillus rivalis*, lateral (20 mm), and (h) *Histiostoma feroniarum hypopus*, ventral (0.2 mm) (CIWEM, 2000). Reproduced with permission of Elsevier Publishing.

*Histiostoma feroniarum*, were almost ubiquitous. All the 24 most common species grazed the biofilm with one exception — the predatory mite *Platyseius italicus* (Learner and Chawner, 1998). Unusual invertebrates are also occasionally recorded. For example, seawater can enter damaged sewers at coastal sites, which allows estuarine and marine species to colonize filters. Jones and Wigham (1993) studied a population of the supralittoral amphipod *Gammarus duebeni* that had colonized and thrived in a percolating filter in Cornwall, England.

## 4.4.1 *Annelida*

Only three families of the phylum Annelida — the Enchytraeidae, Lumbricidae and the Aeolosomatidae — are common in FFRs.

**Table 4.7.** The principal macroinvertebrate species occurring in the surface medium samples of 59 filter beds that were sampled in spring and autumn by Learner (1975a). The average abundance of species (number per $L^{-1}$ of medium) and percentage occurrence are given (Learner and Chawner, 1998). Reproduced with permission of John Wiley and Sons Inc.

| Species | Average abundance (No. $L^{-1}$) | Occurrence (%) |
|---|---|---|
| *Lumbricillus rivalis* | 6055 | 91 |
| *Psychoda alternata* | 3509 | 89 |
| *Histiostoma feroniarum* | 588 | 92 |
| *Pristina idrensis* | 314 | 48 |
| *Bryocamptus pygmaeus* | 146 | 61 |
| *Metriocnemus eurynotus* | 119 | 54 |
| *Enchytraeus coronatus* | 118 | 22 |
| *Platyseius italicus* | 113 | 85 |
| *Psychoda albipennis* | 109 | 72 |
| *Limnophyes minimus* | 98 | 61 |
| *Nais elinguis* | 88 | 52 |
| *Hypogastrura viatica* | 84 | 51 |
| *Histiogaster carpio* | 65 | 66 |
| *Enchytraeus buchholzi* | 47 | 57 |
| *Paracyclops fimbriatus* | 41 | 73 |
| *Chaetocladius perennis* | 31 | 24 |
| *Sylvicola fenestralis* | 25 | 49 |
| *Nais variabilis* | 21 | 54 |
| *Paracyclops chiltoni* | 8 | 18 |
| *Pristina aequiseta* | 8 | 42 |
| *Paracyclops poppei* | 5 | 12 |
| *Eiseniella tetraedra* | 3 | 52 |
| *Tomocerus minor* | 3 | 28 |
| *Dendrodrilus subrubicundus* | 3 | 43 |

*Aeolosoma hemprichi* is common in extended aeration activated sludge systems, but are also found in a number of submerged FFRs where they can become problematic, especially in IFAS reactors (Zhan *et al.*, 2013). They are occasionally recorded in percolating filters, especially in ponded areas. *Aeolosoma hemprichi* has been widely employed in sludge reduction (Liang *et al.*, 2006) and is a highly effective grazer of biofilms. The other two families are almost exclusively found in percolating filters because they are not active swimmers. In contrast, members of the Naididae and Lumbriculidae are more frequently recorded in submerged FFRs as they are active swimmers, although they too are also occasionally reported from percolating filters.

The most frequently observed enchytraeid is *Lumbricillus rivalis*, which was found in 91% of the filters surveyed by Learner (1975a). This species can make up 95–100% of the total enchytraeid population in filters, with maximum population densities reaching 10,000 per liter of mineral medium (Gray, 1980). *Enchytraeus buchholzi* is the next most frequently observed species, being recorded in low numbers in 57% of the filters examined by Learner. Gray observed that *L. rivalis* was distributed throughout the depth of his pilot filters at the low-rate loading, reaching maximum abundance in the central and lower areas of the filters, with the enchytraeids becoming restricted to the lower half of the filters at the high-rate loading. Previous workers had found *L. rivalis* principally in the upper portions of the filters near the surface (Reynoldson, 1947; Solbe *et al.*, 1967; Williams *et al.*, 1969). Maximum numbers of enchytraeids are found during February with minimum densities occurring in August, indicating that the population density is directly linked to biofilm accumulation. However, Enchytraeidae are restricted by the presence of the Psychodid larvae. They reach maximum population densities the month preceding the occurrence of the maximum number of Psychodid larvae, but both reaching maximum abundances in response to the heavy biofilm accumulation. The *Psychoda* spp. and enchytraeids are in direct competition for food and space in the filter, with the larvae being more successful because of their size and rapid growth rate. This increased competition from the larvae causes a dramatic decline in the numbers of enchytraeids present (Gray, 1980). Enchytraeidae have also been found to dominate the grazing fauna under reduced competition from the fly larvae caused by higher hydraulic loadings or lower temperatures (<10°C) (Tomlinson and Hall, 1950; Hawkes, 1955; Solbe *et al.*, 1974). Hawkes (1955) demonstrated in his experiments on dosing frequencies that *L. rivalis* could withstand relatively high hydraulic flows because it possessed strong, curved setae. They could also withstand periods of even higher flow-rates as the cocoons were firmly attached to the medium. However, Enchytraeidae are restricted at higher flow rates (>2 $m^3m^{-3}d^{-1}$) with large numbers lost in the effluent, especially during sloughing (Reynoldson, 1941; 1948), and are generally absent from high-rate filters using either plastic or mineral media at loading rates of >6 $m^3m^{-3}d^{-1}$ (Bruce and Merkens, 1970).

*Lumbricillus rivalis* is capable of increasing its population density rapidly because of the large number of cocoons present in the filter at specific periods and its high population growth rate (Learner, 1972). This allows the Enchytraeidae to respond quickly to changes in the biofilm accumulation and also to changes in the community structure. The number of cocoons in the filter bed is controlled by the biofilm accumulation. More are retained as the biofilm increases, either by adhesion to the biofilm or some other suitable substrate and by being mechanically filtered out of the sewage by the biofilm and humus. At times of low biofilm accumulation, there is a corresponding low abundance of cocoons that are mainly adhering to the actual surface of the media. Reynoldson (1947) and Solbe *et al.* (1967) both found that the abundance of cocoons followed a seasonal pattern of reaching maximum numbers in spring and autumn. In his pilot filters, Gray found that the seasonal abundance of cocoons followed a similar pattern to that of the adult enchytraeids, reaching maximum numbers from February to April and minimum numbers during the summer months after sloughing. Both the adult enchytraeids and the cocoons are eaten by a number of predatory dipteran larvae including *Limnophyes minimus, Metriocnemus eurynotus* and *Psychoda albipennis* (Lloyd, 1945).

Only three lumbricids are frequently recorded in percolating filters: *Eisenia foetida* (31% of the filters surveyed by Learner), *Dendrobaena subrubicunda* (43%) and *Eiseniella tetraedra* (52%). Considerable information regarding these species in filters has been gathered (Tomlinson, 1946a; Hawkes, 1963; Solbe *et al.*, 1967; Solbe, 1971) and this has been reviewed in detail by Solbe (1975). Of the two most common species, *E. tetraedra* is an amphibious species, whereas *D. subrubicunda* is terrestrial in nature; it is commonly found in compost heaps (Gerard, 1964). Therefore, it is not surprising that *E. tetraedra* is more successful at higher hydraulic loadings than the other species. Lumbricids tend to be more numerous in smaller media (Terry, 1951), with worms rarely found near the surface in filters containing larger media or media with a high voidage due to the flushing action of the applied wastewater. Solbe (1971), while examining the depth distribution of lubricids, reported that the population density of *D. subrubicunda* increased toward the base of the filters, whereas *E. tetraedra* is found in the middle regions of the bed. Therefore, the distribution of the lumbricids is apparently related to the hydraulic flow and the size of the interstices.

The presence of lumbricids is generally accepted as an indication that a filter has matured and that the community structure has become stable. However, during high rainfall, large numbers of worms are flushed out of the soil on to paved areas and are washed away in the surface run-off. Thus, where combined sewerage systems are employed, these worms eventually arrive at the treatment plant. Smaller individuals or cocoons not removed by primary sedimentation will become established in the filter, with large population densities being established quickly. *Dendrobaena subrubicunda*, for example, can ingest biofilm at a rate of 133 mg $g^{-1}d^{-1}$ at 15°C, with the entire lumbricid population of a mature filter capable of ingesting an amount of biofilm equivalent to 55% of the daily input of carbonaceous material to the filter (Solbe, 1975).

Annelids, therefore, prevent the accumulation of excess solids and may also be the prime cause of the sloughing of biofilm in some filters during spring. Annelids ingest a wide range of substances, including those not readily degraded by the filter bacteria. Thus, Annelida can reduce the organic matter directly by absorption and indirectly by the formation of excreted aggregates (feces), which readily settle in the humus tank. Although some heterotrophic bacteria flourish in the gut of certain annelids and are passed out with the aggregates, they are also responsible for the consumption of many pathogenic microorganisms. The respiration of the phylum accounts for a small but useful proportion of the carbon dissipated by the filter, which in some filters may be as high as 8.5%. The group causes little nuisance in the filter, although the accumulation of large numbers on the surface can cause alarm to operators even though they are quite harmless. If large numbers are washed out into the humus tank they rapidly die and putrefy, causing a reduction in the dissolved oxygen concentration that can cause problems. They do, however, compete directly with fly larvae and so reduce the potential nuisance caused when adult flies emerge from the filters. A useful identification key has been produced by Brinkhurst (1971).

### 4.4.2 *Insecta*

The members of the various orders of Insecta are principally associated with percolating filters, rarely being found in other kinds of FFRs, or at other stages of wastewater treatment. Many species lists have been compiled (Lloyd, 1945; Tomlinson, 1946a; Terry, 1951; Hawkes, 1963; Solbe *et al.*, 1967). However, in Learner's (1975a) comprehensive survey into the fauna of percolating filters, 186 species of insects belonging to 38 families were recorded (Learner, 1975b).

*Collembola*

Sixteen species of Collembola (springtails) have been recorded from percolating filters, the most important species being *Hypogastrura viatica* (=*Achorutes subviaticus*), which was recorded in 57% of the filters surveyed by Learner (1975b), with *Tomocerus minor* (28%) and *Proisotoma minuta* (21%) also frequently observed. However, only *H. viatica* are found in large numbers, reaching maximum densities of 3,800 individuals per liter of medium. They are active grazers found throughout the depth of the filter and feed on a wide variety of organic material including fungal hyphae and spores, bacteria, dead and decaying plant material, and algae (Learner, 1975b). *Hypogastrura viatica* is extremely sensitive to increased hydraulic loadings (Hawkes and Jenkins, 1955, 1958; Wheatley, 1976). For example, Tomlinson and Hall (1950) recorded the maximum abundance of the springtail at 1.5 $m^3m^{-3}d^{-1}$, with none found at loads of >3.0 $m^3m^{-3}d^{-1}$. *Isotoma olivaceaviolacea* has a preference for drier areas, such as inter-jet zones in conventional filters (Hawkes, 1959). Gray (1980) found *I. olivaceaviolacea* more abundant in random plastic than in mineral media because each piece of random plastic media, such as Flocor RC or RS, has a dry area.

### *Coleoptera*

Beetles (Coleoptera) are not common in percolating filters. Only members of the Hydrophilidae and Staphylinidae are frequently encountered, with the most widely occurring species being *Cercyon ustulatus* and *Platystethus arenarius*, which were both observed by Learner during his survey. Coleoptera are rarely found as larvae in the filter and presumably do not breed within the filter environment.

### *Diptera*

The most important insects found in filters are the Diptera, with 28 species from 11 dipteran families regularly occurring (Learner's survey). The principal filter species belong to the genera Psychodidae, Chironomidae and Anisopodidae, all of which are able to breed successfully within the filter bed, with the larvae being abundant and feeding on the biofilm. Many more species of Diptera are associated with filters and are captured in very large numbers on insect traps set on, or near to, filters. All the species that are able to breed in the filter environment have an important role in the actual purification process.

The dipteran larvae and Collembola are very active grazers, thus preventing the filters from becoming blocked with excessive biofilm accumulation (Williams and Taylor, 1968). They are most active during the warmer months, although large populations of some species, such as *Sylvicola*, are still active even during winter. The function of the Insecta in the purification process is the same as that of the Annelida; they digest the biofilm and absorb some of the organic matter, and form dense fecal pellets that rapidly settle. Williams and Taylor (1968) demonstrated that solids in the effluent from filters containing *Psychoda* and/or *Lumbricillus* spp. settled far more rapidly, with 65–70% of the total settleable solids settling within an hour compared to 34% in the control filter without macroinvertebrates. The sludge from the filters containing macroinvertebrates contained a high density of animal fragments and fecal pellets.

Three species of Psychodidae occur in filters. *Psychoda alternata* is the most frequently observed species, occurring in 89% of the filters sampled by Learner and reaching maximum densities of 44,700 $L^{-1}$ of medium (Fig. 4.13). Therefore, this species, together with *Lumbricillus rivalis*, must be considered the major grazers in percolating filters. *Psychoda albipennis* (formally *P. severini*) is also widely distributed (72%) but reaches lower densities (1,240 $L^{-1}$) than *P. alternata*, whereas the third species, *Psychoda cinera*, is found only in 4% of the filters and in low numbers (Fig. 4.14). *Psychoda albipennis* is parthenogenetic. Like *P. alternata*, it can carry out its entire life cycle within the filter bed. It is therefore quite common to find adult flies deep within the medium. This is in contrast to many other dipterans where the adult flies need to leave the filter after pupation to swarm and mate before returning to the filter environment to lay eggs. These two species do not actively compete, as *P. albipennis* can reproduce at temperatures below 10°C and is more abundant than *P. alternata* during the winter and spring, whereas the latter is dominant in summer. The abundance of *P. alternata* is positively correlated with temperature. Learner (1975*a*) clearly illustrated that the reproductive potential (life cycle) of *P. alternata* is controlled by temperature and that it had the most rapid development rate at temperatures exceeding 10°C of any of the insects found in the filter environment. Obviously, there is a lag phase between maximum food availability and the resultant increase in the number of grazers. In the case of *P. alternata*, this took 1 to 2 months, depending on the filter temperature (Solbe and Tozer, 1971). This explains why Gray (1980) did not observe any direct correlation between biofilm weight and density of the Psychodid larvae. The restriction in the surface accumulation of *Psychoda* spp. is due partly to its sensitivity to the hydraulic flow (Tomlinson and Hall, 1950; Hawkes, 1955; Lumb and Eastwood, 1958). The effect of various distribution systems was studied by

Hawkes (1959), who found that splash plates produced an even distribution of sewage resulting in a large accumulation of biofilm and a high density of *Psychoda* in the top 600 mm. By increasing the velocity of application, the *Psychoda* populations were reduced in the surface layers and were forced below 600 mm. This resulted in more biofilm in the top layer. Tomlinson and Hall (1950) found that the abundance of the *Psychoda* larvae decreased if the hydraulic load exceeded 3.6 $m^3m^{-3}d^{-1}$ even though the biofilm was thick. In their experimental filters, Bruce and Merkens (1970) found that large numbers of

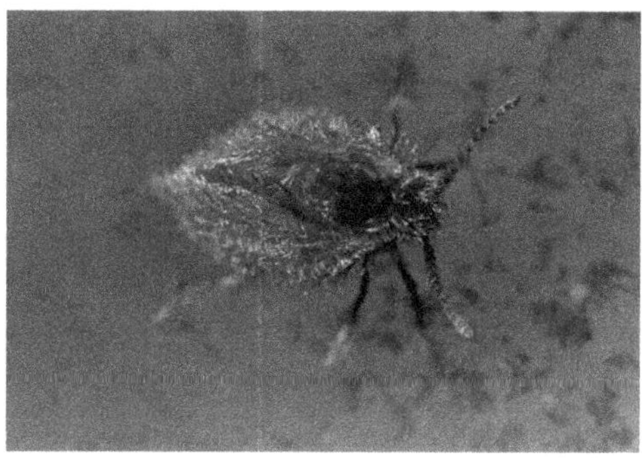

(a)

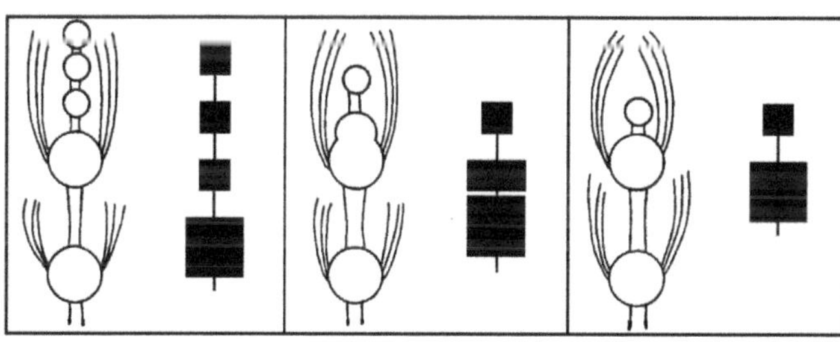

(b)

**Fig. 4.14.** The dipteran most associated with percolating filters are small moth-like flies less than 5 mm long of the genera Psychoda. (a) *Psychoda alternata* is extremely common and is known as the "sewage filter fly". It is readily distinguished from the other two species of Psychoda found in filters by dark tufts of hairs at the tips of some wing veins. *Psychoda cinerea* and *Psychoda albipennis* have hair-less wings. (b) Flies can be positively identified by low-power microscopic examination of their antennae. The *Psychoda cinera* terminates in three small segments, while the *P. alternata* has two large segments that are broadly joined and a smaller terminal segment. The *P. albipennis* is a single small segment.

*Psychoda* larvae were present in filters loaded at 6.0 $m^3m^{-3}d^{-1}$. This was probably due to the higher voidage as it has been shown that smaller media restrict the natural life cycle of the insects (Tomlinson and Stride, 1945; Hawkes and Jenkins, 1951; 1955). Using multivariate techniques, Learner and Chawner (1998) were able to reanalyze the unique data set of invertebrates in British percolating filters collected by Learner (1975a). Using the TWINSPAN multivariate analysis, they found that *P. albipennis*, along with the copepod *Bryocamptus pygmaeus*, were indicators of low organic loading, whereas abundant populations (>1,000 larvae $L^{-1}$ of medium) of *P. alternata* indicated high organic loadings.

Three chironomid species were frequently isolated by Learner (1975a): *Chaetocladius perennis* (formally *Hydrobaenus perennis*) (24%), *Metriocnemus eurynotus* (formally *Metriocnemus hygropectricus*) (54%), and *Limnophyes minimus* (formally *Hydrobaenus minimus*) (61%). They were all able to breed in filters with maximum larval densities reaching 1,400, 1,892 and 3,460 $L^{-1}$ medium, respectively. The larvae of *C. perennis* migrate in a generally downward direction during their development. This results in the maximum abundance of larvae being found at the base of filters, with high numbers of larvae being washed out in the final effluent and a scarcity of pupae in comparison with other species. The final larval instar finally burrows deep into the biofilm to pupate (Lloyd *et al.*, 1940). Like other chironomids, the species is common in all types of filter media and is abundant from February to August, although it is more readily flushed out of smoother faced mediums. Chironomid larvae are generally only found in large numbers in lightly loaded filters (Tomlinson and Stride, 1945) when the biofilm accumulation is thin (Terry, 1956; Hawkes and Shepherd, 1972). *Metriocnemus eurynotus* is found in the greatest numbers during autumn. Due to the upward migration of the species before pupation (Dyson and Lloyd, 1936), the larvae are found in large quantities in the upper section of filters. *Limnophyes minimus* is restricted to low-rate filters. It is found from May to October throughout the filter and is most successful when the biofilm is thin. Chironomid larvae are only found in relatively small numbers in filters and have a minor role in the purification process. However, the larvae are capable of successfully competing with the other macroinvertebrates of the filter under favorable conditions. Lloyd *et al.* (1940) reported that chironomid larvae were able to compete with psychodid larvae for available food, reducing the population densities of *Psychoda* spp. at the surface of the filter and causing the extension of the species distribution deeper into the filter. In heavy biofilm conditions, the *Psychoda* and *Sylvicola* larvae are more able to cope than the larvae of the chironomid species. They utilize their respiratory siphons when buried in the thick layer of biofilm. Both *L. minimus* and *M. eurynotus*, like *P. albipennis*, have shorter life cycles than *P. alternata* at the lower range of temperatures recorded in filters. This may account for their relative success in the areas of the bed most affected by exposure to the air and is frequently cold in comparison to the other areas of the filter. Examples of vertical and seasonal variations of some grazing organisms, including *Psychoda* spp. and *C. perennis* larvae, are shown in Fig. 4.15.

*Sylvicola* (formally *Anisopus*) *fenestralis* is the only member of the Anisopodidae recovered from filters being widely distributed (49% of filters) and in large numbers (1,680 $L^{-1}$ medium). The adult fly has extremely large wings in comparison with other filter flies and is a powerful insect with large larvae. Like *Psychoda*, it can complete its life cycle within the filter bed with maximum numbers occurring between April and June. The larvae are generally found in the largest numbers in the top 600 mm of filters irrespective of the organic or hydraulic loading (Tomlinson and Hall, 1950; Gray, 1980) (Table 4.8), while the pupae reach maximum densities between 300–900 mm depth. Hawkes (1963) found that the vertical distribution of the species could be altered by high instantaneous rates of domestic wastewater. *Sylvicola fenestralis* abundance is closely and directly related to biofilm distribution (Hawkes, 1952). However, recent studies suggest that the amount of biofilm is not always an important factor determining the vertical distribution of this species. This species was reported as being more successful in random plastic rather than slag medium by Gray (1980). Each module of random plastic media has some of its surface area free from the biofilm, which is often quite dry, and this may be the reason why the pupae, in particular, were found in comparatively large numbers in this filter. *Sylvicola fenestralis* requires a drier environment for successful pupation than that

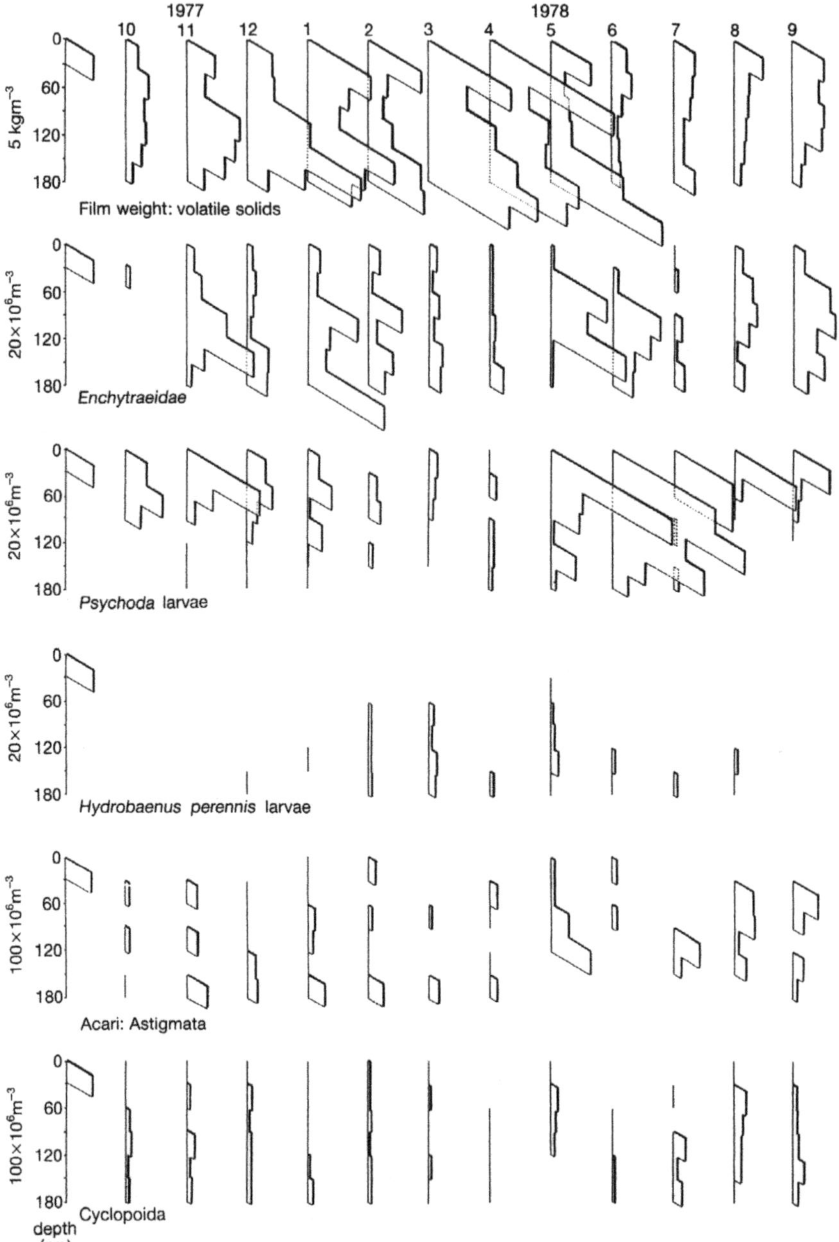


**Fig. 4.15.** Vertical distribution of film and macrofauna in a low-rate filter containing 50 mm blast-furnace slag medium.

**Table 4.8.** Vertical distribution of common insects at the Minworth Wastewater Treatment Works (Filter Block B) in Birmingham during August 1968 (Learner, 1975a). A dash is equivalent to zero.

| Depth (m) | No. per liter of medium | | | | | | |
|---|---|---|---|---|---|---|---|
| | *Anurida granaria* | *Hypogastrura viatica and H. purpurescenes* | *Cercyon ustulatus* | *Sylvicola fenestralis* | *Psychoda alternata* | *P. cinerea* | *P. albipennis* |
| 0.00–0.15 | 18 | 18 | — | — | 89 | 89 | — |
| 0.15–0.30 | 71 | 18 | 36 | 107 | 36 | 214 | 53 |
| 0.30–0.46 | 142 | 89 | 36 | 18 | 18 | 231 | 18 |
| 0.46–0.61 | 1175 | 53 | 53 | 71 | — | 303 | 18 |
| 0.61–0.76 | 552 | 125 | 18 | — | 36 | 71 | — |
| 0.76–0.91 | 445 | 71 | — | — | — | 89 | — |
| 0.91–1.07 | 677 | 142 | 18 | — | — | 36 | — |
| 1.07–1.22 | 944 | 71 | — | — | 18 | 36 | 18 |
| 1.22–1.37 | 427 | 53 | — | — | — | 53 | 18 |
| 1.37–1.52 | 374 | — | — | — | — | 18 | — |
| 1.52–1.68 | 214 | 71 | — | — | — | — | — |
| 1.68–1.83 | 641 | 89 | — | — | — | — | — |

tolerated by the larvae (Hawkes, 1952), and this is thought to be the reason why the larvae are reported to migrate to drier areas in conventional plants (Learner, 1975b). *Sylvicola fenestralis* appears unaffected by competition with either chironomid or psychodid larvae.

Many of the dipteran flies form dense swarms as they emerge from the filter bed. For some chironomids swarming is a prelude to mating (Fig. 4.16). The number of flies emerging can be so large that from a distance, it has the appearance of smoke coming from the reactor surface. Emergence is affected by temperature, light intensity and wind velocity. Learner (2000) found that the average air temperature was a major environmental factor determining the phenology of filter flies, allowing emergence patterns to be predicted. Insects will not fly unless the temperature is above a threshold value, for example, 10°C for *P. alternata*, 7–8°C for *L. minimus*, and 4°C for both *Metriocnemus* spp. and *S. fenestralis* (Learner, 1975b). Hawkes (1961) found that for every 1.2°C rise in temperature above the threshold value for up to 24°C, the number of *S. fenestralis* flies in flight doubled.

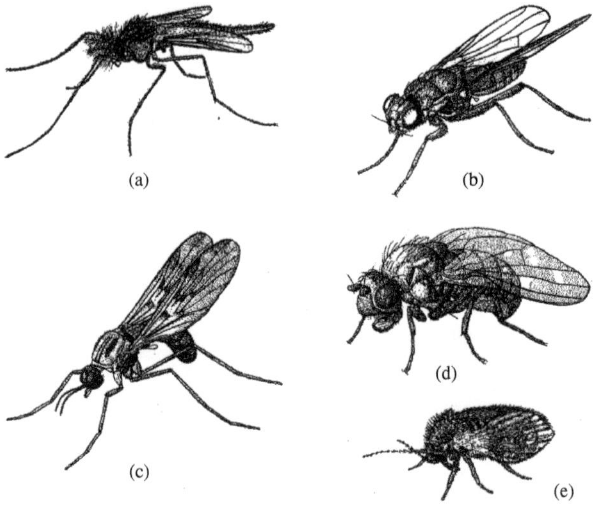

**Fig. 4.16.** The major species of flies associated with percolating filters: (a) *Metrioconemus hygropectricus*, (b) *Spathiophora hydromyzina*, (c) *Sylvicola fenestralis*, (d) *Scatella silacea*, and (e) *Psychoda alternata* (IWPC, 1988). Reproduced with permission of the Chartered Institution of Water and Environmental Management.

*Chaetocladius perennis* and *P. albipennis* are cold-adapted species that emerge principally in the spring. *Psychoda alternata* and *P. cinera* are warm-adapted species that emerge mainly in the summer, while *L. minimus* emerges principally in late summer to early autumn. *Metriocnemus eurynotus* is variable in its time of emergence, with other environmental variables besides temperature being important (Learner, 2000). Diel periodicity is observed for all the species with the peak of emergence for *P. alternata* and *P. albipennis* in the early afternoon, and *S. fenestralis* at dusk with a smaller peak at dawn. Wind velocity is an important factor with swarms quickly broken up and individuals unable to fly if the wind is too strong. In essence, the stronger the fly the greater the wind velocity it can withstand, with *S. fenestralis* able to withstand velocities of up to 6.7 m $s^{-1}$ (Hawkes, 1952; 1961). At the treatment plant, the density of flies can be problematic, which makes working conditions unpleasant as flies are drawn into the mouth and nostrils, as well as being caught in the eyes. None of the commonly occurring flies, including *S. fenestralis*, bite. However, although harmless, the large size and intimidating appearance of *S. fenestralis* can alarm people, which result in frequent complaints from nearby residents. Psychodid flies are found up to 1.6 km away from the treatment plant, although this depends on the direction of the prevailing wind. *Sylvicola fenestralis* are found in large numbers up to 1.2 km, although few reach further than 2.4 km from the plant. However, flies are well known to cause both aesthetic and public health problems to those living or working close to treatment plants (Woods *et al.*, 1978; Painter, 1980; Palfrey *et al.*, 1992), with *P. alternata, P. albipennis, S. fenestralis, L. minimus* and *M. eurynotus* being the main nuisance species. However, Learner (1975a) found that *P. alternata* comprised 80% or more of the total annual emergence of flies from 10 out of the 17 filters where emergence traps were located. Details of the less common dipterans caught from fly traps (Fig. 4.17) placed on the surface of filter beds from 10 wastewater treatment works around the UK by Learner (2000) are summarized in Table 4.9. June and July are the principal months of emergence.

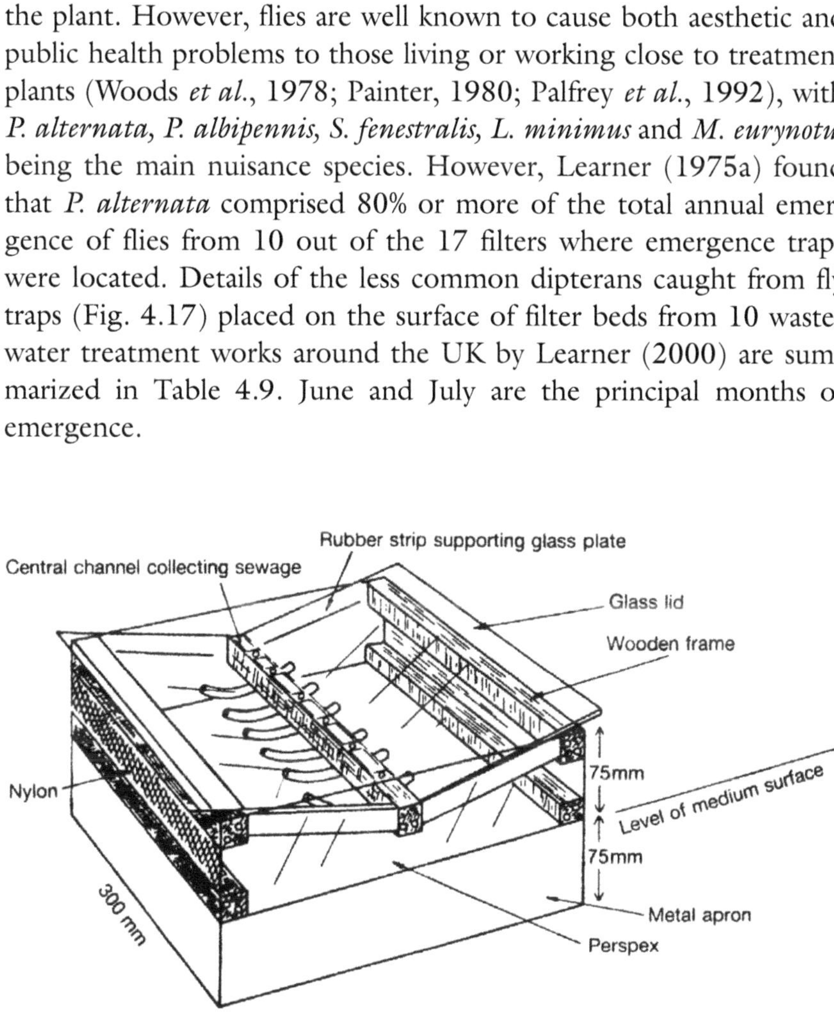


**Fig. 4.17.** Insect trap for determining the emergence of insects from the surface of percolating filters. The inclined glass or plastic sheets are covered with an adhesive and placed with their sticky side downward so that the emerging insects become trapped (Solbe *et al.*, 1967). Reproduced with permission of the Chartered Institution of Water and Environmental Management.

**Table 4.9.** Monthly occurrence of less common insect species caught on fly traps set by Learner (2000). Double crosses indicate periods when the highest numbers were trapped. Reproduced with permission of Elsevier Publishing.

| Species | January | February | March | April | May | June | July | August | September | October | November | December |
|---|---|---|---|---|---|---|---|---|---|---|---|---|
| *Trichocera maculipennis* | | + | ++ | + | + | + | | | | + | + | |
| *Sylvicola fenestralis* | | | | + | + | + | ++ | ++ | + | + | + | + |
| *Metriocnemus hirticollis* | + | + | + | + | + | + | + | + | + | ++ | + | |
| *Coproica vagans* | | | | | + | ++ | ++ | + | + | | | |
| *Elachisoma aterrimum* | | | | | | ++ | + | + | + | | | |
| *Halidayina spinipennis* | | | | + | + | ++ | ++ | + | + | + | | |
| *Ischiolepta pusilla* | | | | + | | ++ | + | + | + | | | |
| *Leptocera caenosa* | | | + | + | + | ++ | ++ | + | + | + | + | |
| *L. fontinalis* | | | + | + | | ++ | | | | | | |
| *L. limosa* | | | | | | | + | ++ | | | | |
| *L. nigra* | | | | + | | + | ++ | ++ | + | + | + | |
| *Limosina silvatica* | | | | | | | | | | ++ | + | |
| *Opacifrons coxata* | | | | | + | + | ++ | + | | + | | |
| *O. humida* | | | | | | ++ | + | + | + | | | |

| Species | January | February | March | April | May | June | July | August | September | October | November | December |
|---|---|---|---|---|---|---|---|---|---|---|---|---|
| *Opalimosina liliputana* | | | | | | ++ | | + | + | + | | |
| *O. mirabilis* | | | | | | ++ | + | + | + | + | | |
| *Spelobia bifrons* | | | | | | ++ | + | + | | | | |
| *S. cambrica* | | | | | + | ++ | + | + | ++ | | | |
| *Sphaerocera curvipes* | | | | + | | ++ | + | | | | | |
| *Trachyopella atomus* | | | | | | ++ | + | + | + | + | | |
| *Scatella silacea* | | | | + | + | + | + | ++ | + | ++ | + | |
| *S. stagnalis* | | | | | | + | ++ | | + | + | | |
| *Culicoides vexans* | | | | | | | | ++ | | | | |
| *Themira putris* | | | | | + | + | + | ++ | | | | |

### *Control options for fly emergence*

Three control options are available for remedying fly nuisance: physical, chemical and biological. Physical methods have been least successful and so are not generally recommended. In the United States (US), trickling filters are flooded to eliminate nuisance fly species. This is achieved by blocking the outlet and allowing the influent to fill the bed until the filter medium is completely submerged. The bed is left for 24 hours, during which time the adult flies drown, and the larvae and pupae come to the surface where they can be skimmed off. This approach damages the ecology of the system and adversely affects performance, providing at best a temporary reduction in the number of flies emerging (Otter, 1966). Chlorination is also used in the US but severely damages all the biota, leading to a significant loss in performance. Covering the surface of the medium with a layer of finer media (13–19 mm) to a depth of 250 mm reduces the numbers of adult *Psychoda* and *Sylvicola* emerging, but results in severe ponding during winter. Enclosing filters can be effective but rather expensive. However, they are widely adopted in industrial units, where ventilation and bed temperature are also controlled. However, Painter (1980) suggests that this remedy may not always work and quotes an example when after covering a filter, the extremes in temperature were reduced and led to an overall increase in fly production, with flies escaping via ventilation ports. The most widely adopted control method for controlling filter flies in the UK has been the use of chemical insecticides. Two insecticides, Actellic M20 (pirimiphos methyl) and Dimilin (diflubenzuran), which are O-2-diethyla-mine-6-methylpyrimidin-4-yl-O-O-dimethyl phosphorbthioate and 1-(4-chlorophenyl)-3-(2,6-difluorobenzoyl) urea, respectively, have been widely used. However, neither insecticide appears particularly effective against *S. fenestralis.* These pesticides have since lost clearance for use on filters within the European Union, including the UK, due to persistence in the environment, as well as concerns about resistance developing within fly populations.

Current control methods use entomopathogenic bacteria (Houston *et al.*, 1989a; 1989b; Coombs *et al.*, 1991) or insect growth hormones (Kamei *et al.*, 1993, Coombs *et al.*, 1996) against the larvae. The most widely adopted technique is the use of the entomopathogenic bacterium *Bacillus thuringiensis* var. *israelensis*. The bacterium, available as commercial preparations known as Teknor® and VectoBac® 12AS, is effective against fly larvae only and does not affect other filter fauna or performance. The bacterium produces crystalline cell inclusions during sporulation that are highly poisonous when eaten by the dipteran larvae. Application rates are dependent on the target species and size of the larvae, with chironomid larvae less susceptible than *Psychoda* larvae. Between 10–20 mg of the active ingredient is added to a liter of wastewater, which is equivalent to 0.8–1.7 ml of the undiluted product per liter of settled sewage applied to the filter for 30 minutes, which is repeated as required. It takes between 7 and 14 days for a reduction in emergence to become apparent. Wenjere (2017) compared the effectiveness of VectoBac® 12AS at three different recommended concentrations — 80, 160 and 230 mg $L^{-1}$ — for the control of filter flies. Two treatments at 80 mg $L^{-1}$ effectively controlled *P. alternata*, achieving a reduction of larval and adults within 48 hours of 93%, which continued for the next 23 days of the test. A 100% reduction of *S. fenestralis* was achieved with a single treatment at 160 mg $L^{-1}$ within 2 days and lasting for the next 14 days of the trial. Two treatments at all dosing levels gave a 100% reduction of *L. minimus* after 48 hours. However, control fell to 30% one month later at the low dose but remained high at the medium (91%) and high (90%) dose rates. The study shows that careful trials are required to determine the optimum control strategy for any particular filter and nuisance species.

Insect growth hormones inhibit the development of insect growth stages and have been successfully employed in controlling midge emergence in urban areas from retention-detention ponds, as well as in trickling filters in the US. Normally, insects produce hormones to regulate their development by the inhibitory metamorphosis of the larval instars. A reduction of the hormone causes pupation and its absence results in the emergence of the adult (imago). The addition of the hormone analog permits the development of the larvae but prevents pupation or emergence, thus retaining maximum grazing activity but preventing fly emergence (Coombs *et al.*, 1996).

Pyriproxifen has been widely used to control *Psychoda*. In the US, the most widely used insect growth regulator for trickling filters is Stike®, which is produced by Zoëcan. It is initially used at a rate of 39 mg of water-soluble reagent per cubic meter of wastewater for two weeks, then reduced to between 19.5 to 23.0 mg $m^{-3}$ to sustain its action. Due to the low persistence of both of these new methods and their cost, they should only be used when a major egression of flies is expected which requires accurate prediction (Learner, 2000).

Routinely, limited control can be obtained by changing the operational practice of the plant, for example, by limiting the amount of biofilm accumulation, especially at the surface. This can be achieved by reducing the f/m ratio or by using one of the modifications such as recirculation, double filtration or alternating double filtration. Increasing the hydraulic loading may also make it more difficult for dipterans to complete their life cycles in the filter. Continuous dosing using nozzles prevents their emergence and severely reduces the available surface area for the flies to lay eggs.

### *Acari*

Astigmata mites (Acari) are extremely abundant in percolating filters and are associated with drier areas of the medium, especially the dry inter-jet zone of filters and random plastic medium where each module supports a dry niche. Maximum abundance occurs in spring, with maximum densities in slag media reaching 3,900 $L^{-1}$, and at almost an order of magnitude higher at 32,000 $L^{-1}$ in random plastic media (Gray, 1980). They are more abundant in low-rate systems but are found in reduced numbers in high-rate filters, provided a suitable niche is available. Four species are particularly common: *Histogaster carpio, Histiostoma feroniarum, Rhizoglyphus echinopus* and the predatory mesostigmata *Platyseius italicus* (Baker, 1961; 1975). The Astigmata can respond quickly to increases in biofilm accumulation, as opposed to other macrograzers. The reproductive potential of the Astigmata and, in particular, *H. feroniarum*, is far greater than that of the dipteran larvae (Learner, 1975b) or the Enchytraeidae (Learner, 1972). Hughes (1961) found that *H. feroniarum* completed its life cycle in only 2 to 4 days at 20–25°C, whereas *R. echinopus* took 9 to 13 days over a similar temperature range (Zachvatkin, 1941). Initially, the mites take advantage of the large accumulation of biofilm due to a rapid reproductive rate, but the enormous numbers of Psychodid larvae that will develop subsequently force the mites into other areas of the filter bed or cause a reduction in the total population density. The predatory species *P. italicus* has been shown to feed on a wide variety of invertebrates but mainly on the *L. rivalis*, although it does not eat the cocoons (Baker, 1961). It is closely associated with Enchytraeidae density, often being found together on the surface in large numbers (Tomlinson, 1946a; Gray, 1980).

Not all the macrofauna found in filters are grazing on the biofilm. The spiders and mesostigmatid mites are predators, as are some of the chironomid larvae when the biofilm becomes scarce. Duffey (1997) only recorded two spiders — *Leptorhoptrum robustum* and *Erigone longipalpis* (*Linyphiidae*) — in large numbers in the filter beds at the Minworth Wastewater Treatment Works in Birmingham. Where humus (dead organic matter) accumulates within the filter, saprophytic species such as *Cercyon ustulatas* and the larvae of *Spaziphora hydromyzina* are found. A number of the dipteran larvae are preyed on or parasitized by other flies.

### 4.4.3 *Mollusca*

Land molluscs, especially slugs, are common on the surface of percolating filters, with the common grey field slug *Deroceras reticulatum* (formally *Agriolimax reticulatus*) commonly recorded (Table 4.1). Just as in natural habitats, they feed on organic matter and are especially associated with algal growth when it occurs on surface media, where they may have a role in preventing surface ponding. Aquatic snails, in particular, the *Radix peregra* (formally *Lymnaea peregra*) cause problems when they penetrate the sewer networks, possibly from groundwater infiltration or more likely due to flooding, and become established in percolating filters, especially those containing random plastic media. Both *Omphiscola glabra* and *R. peregra* have also been recorded in the mineral-packed media in the UK but in relatively small numbers. *Omphiscola glabra* is now on the IUN Red List of Threatened Species and classified as *near threatened*. With the increasing loss of marshes and other transitional aquatic-terrestrial habitats, percolating filters can act as an important reservoir for threatened species. In the US, snails of the genera *Physa* and *Physilla* have been widely reported in filters (Learner, 1975). Clay (1964) also reported two slugs (*Arion hortensis* and *Deroceras agrestis)* and five gastropods, including *R. auricularia*, *L. stagnalis* and several from the family Physidae — *Physa cubensis*, *Physa halei* and *Physa integra*.

A severe infestation of *R. peregra* at the Berkhamsted Wastewater Treatment Works significantly reduced the performance of the filters, which was brought under control by using the biocide copper sulfate (Thomas and Dussart, 2003). Problems were also reported on plastic mineral media filters used for ammonia removal due to overgrazing of the biofilm by an infestation of the mollusc, which eliminated the nitrifying bacteria. The solution was the replacement of existing nitrifying filters with biological aerated filters, in which the snails could not survive (Jiang *et al.*, 2003). These species and their shells can also cause problems by blocking drains and pipework and especially by damaging pumps. Snails found in filters are pulmonates, and therefore they breathe air rather than utilize dissolved oxygen through the use of gills, so they do not thrive in the submerged environments of other FFRs.

The ecology of percolating filters has been excellently reviewed by Hawkes (1983b). The physico-chemical nature of the filter bed environment is of prime importance in determining the community structure of each bed. Using TWINSPAN, a powerful multivariate program for ecological spatial data, Learner and Chawner (1998) classified 59 filter beds using species associations (Fig. 4.18). A second program, DECORANA's ordination determined the abiotic variables likely to affect faunal composition (Fig. 4.19). From this study, the abiotic factors most likely to affect faunal composition were the organic loading, biofilm accumulation, air temperature, size of the medium, and the age of the bed. All these variables, except air temperature, are under operator control, giving the potential for the faunal community structure to be modified and even actively managed.

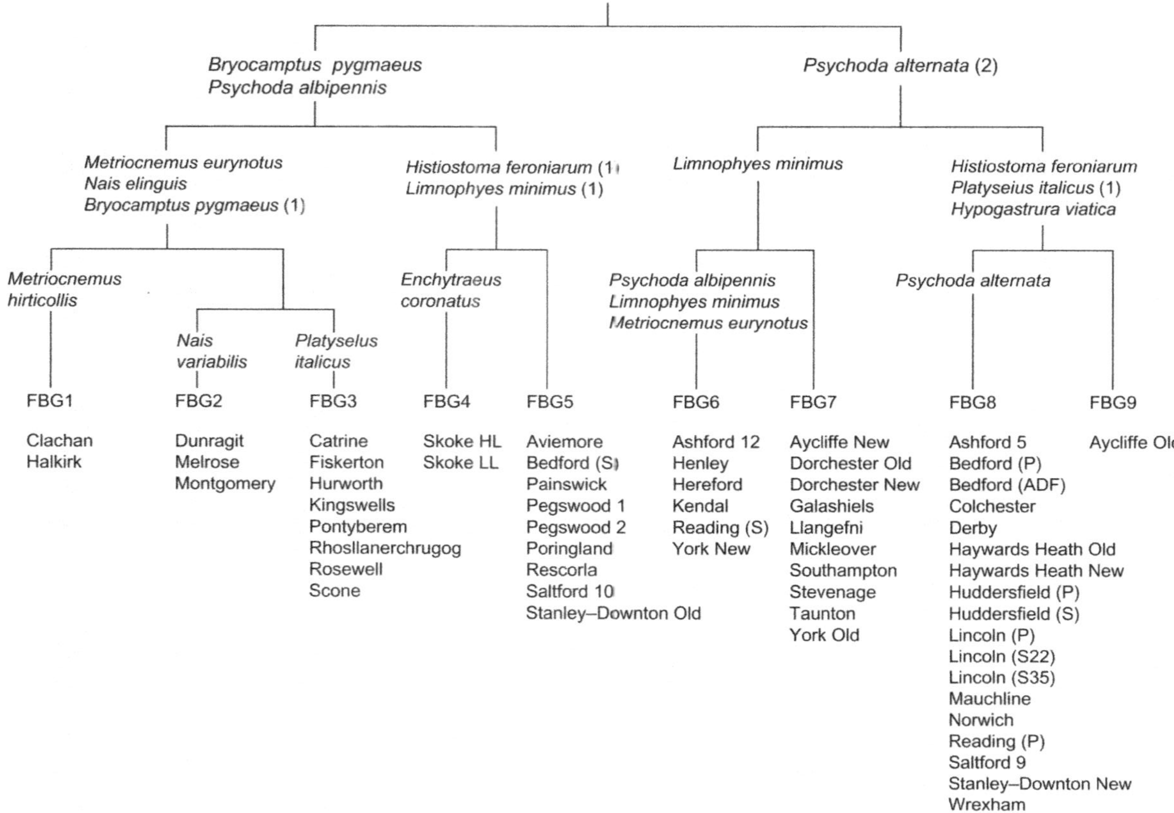


**Fig. 4.18.** TWINSPAN classification of the percolating filter beds surveyed by Learner (1975a). The indicator species for the various divisions are shown. The parentheses next to an indicator species denotes an abundance threshold: 1 = >100 $L^{-1}$ of medium, and 2 = >1000 $L^{-1}$ (Learner and Chawner, 1998). Reproduced with permission of John Wiley and Sons Inc.

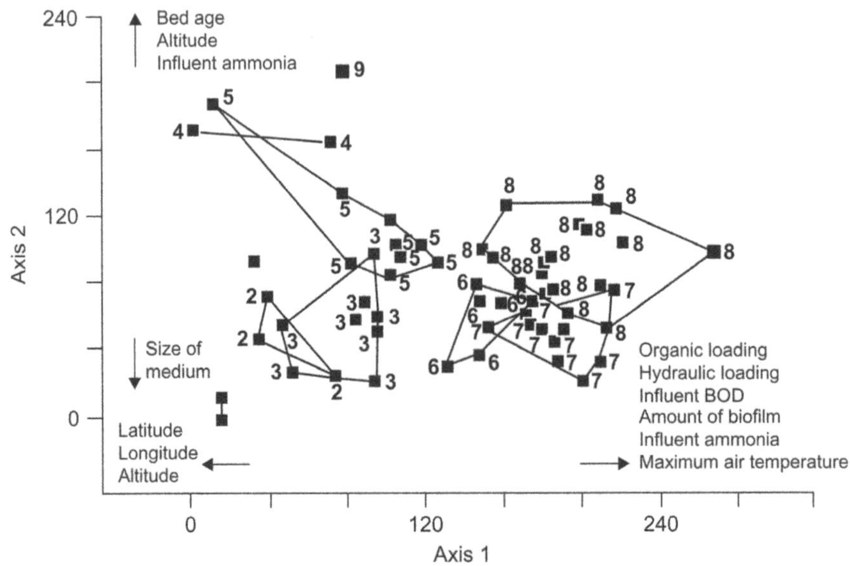


**Fig. 4.19.** DECORANA ordination of the filter beds. The filter-bed group to which each bed belongs, as classified by TWINSPAN (Fig. 4.18), is indicated by the numerals. The environmental variables that correlated with the axes are also shown; the arrows indicate the direction of increasing magnitude (Learner and Chawner, 1998). Reproduced with permission of John Wiley and Sons Inc.

## 4.5 Biological Analysis

Regular biological analysis of the biofilm from FFRs can often provide answers to operational problems as well as being an ideal environment for ecological research (Gray, 1982). The objective of sampling is to obtain a representative sample of the biofilm, which requires pieces of the medium, on which it grows, to be collected and removed from the reactor. Where the medium is fixed *in situ*, the surface of the medium can be sampled by scraping the biofilm off a unit area (e.g., RBCs). Within IFAS systems (Fig 1.2(f)), both the medium and the liquid phase (i.e., the mixed liquor) will need to be sampled and analyzed. Where the medium is in fixed layers, then the community structure may alter vertically or horizontally. For example, in percolating filters, the community structure varies significantly with depth, whereas in RBCs, it varies horizontally between disc sections as the influent moves toward the outlet. Only in completely mixed systems where the strength of the wastewater is the same can it be assumed that there will be a similar biofilm community throughout the reactor, e.g., MBBR. In this section the sampling of percolating or trickling filters is discussed, although the methods can be adapted to all media types and FFR processes.

Learner (1975a) used a single sample of medium collected from the surface of each filter plus a one liter sample of the final effluent for biological analysis during his survey of percolating filters. However, only limited information can be obtained by taking surface samples as the strength of the wastewater falls as it passes through the filter, which affects the community structure of the biofilm. In practice, it is not possible to collect medium from depths exceeding 500 mm by digging a hole in the surface of the bed without causing subsequent distribution problems by forming a short circuit for the influent, thereby reducing overall retention time. Thus, little information can be obtained regarding the vertical distribution of the biofilm or the organisms restricted to the lower half of the filter by this method. Also, where high hydraulic loadings or simple flow-on distributors are used, then little biofilm and a restricted fauna will be present in the top layer of the medium anyway. It is important, therefore, to be able to sample the medium throughout the depth of the filter in such a way as to cause as little disturbance and damage as possible. This is facilitated by the provision of sampling tubes, which allows the vertical distribution of biofilm and fauna to be measured quantitatively and qualitatively.

Sampling tubes are perforated pipes that fit into perforated baskets that contain the medium. The baskets can then be lifted out, allowing the medium and associated biofilm to be sampled at any depth. Gray and Learner (1983) used perforated thick-walled (25 mm) ABS plastic pipes with an internal diameter of 150 mm. The pipes were the same depth as the bed with the top of the plastic pipe just below the surface of the medium. Each pipe was perforated with 38 mm diameter holes some 6 mm apart. Hawkes and Shephard (1972) used smaller holes, less closely spaced, in their sampling columns but an examination of the voids within full-scale percolating filters indicated that larger holes would simulate actual operating conditions more closely. The larger holes also allow greater redistribution of wastewater and movement of loose solids and filter fauna between the sampling column and the surrounding medium. However, the diameter of the perforations was not so large that the blast-furnace slag or random plastic media (50 mm nominal size) used by Gray and Learner would protrude through them sufficiently during settlement to interfere with the removal of the sampling baskets. Each sampling column contained six rigid plastic-coated wire-mesh baskets, each 300 mm long and 146 mm in diameter. Two plastic-coated wires were welded on to each sampling basket to aid their removal. The wires were just thin enough to pass between the baskets and the wall of the sampling column. Biofilm accumulated on these wires and on the pieces of medium that protruded slightly through the holes in the wall of the pipe, thus restricting the free vertical passage of sewage through the gap between the baskets and the pipe. Many other workers have used such removable baskets, utilizing a wide variety of designs and materials (Williams *et al.*, 1969; Hawkes and Shephard, 1972; Wheatley, 1976; Rowlands, 1979; HMSO, 1988) (Fig. 4.20). Horizontal distribution of the

(a) (b)

**Fig. 4.20.** (a) Sampling tubes placed in a filter containing granite medium. (b) A sampling basket removed from a filter containing a random plastic filter medium.

biota can be investigated by using extra sampling baskets housed in shorter sections of perforated pipes positioned in the surface of each filter, thus providing details of the distribution of biofilm in the top 300 mm. This provides useful information on the development of biofilm and the possibility of ponding.

### 4.5.1 *Sampling Procedure*

Removing sampling baskets from their tube will cause some disturbance and so they should be left as long as possible (at least a month) between sampling. The sampling procedure should always be the same. The distribution system must be turned off before sampling, with the baskets removed one at a time and carefully labeled. The basket should be left for five minutes for excess liquid to drain before a subsample of the medium is taken. Gray and Learner (1983) carefully graded the medium within their sample baskets so that four pieces of medium were equivalent to 250 $cm^3$ of medium, and this was taken along with its attached biofilm for analysis. The pieces were removed randomly from different points within the basket to give a representative sample and placed in a labeled plastic bag that was sealed to prevent evaporation and loss of material. The remaining medium was carefully replaced in the baskets in the same order in which they had originally been removed, with the medium from the previous sampling taking the place of the sampled pieces. The sampling procedure should not exceed 20 minutes for each filter to minimize the damage to the rest of the biota. It is important to carefully replace all the pieces of medium, matching up the disturbed surface biofilm so that only a small quantity of biofilm is dislodged and washed away when the distribution system is switched on again. The sample bag should be returned to the laboratory as quickly as possible where it can be weighed to give an estimate of the biomass of biofilm on the medium (Sec. 3.4.5). The addition of preservatives such as formaldehyde or alcohol makes identification of the microfauna, especially the Protozoa, almost impossible. Therefore, it is advisable to store samples at 4°C until they can be processed further.

### 4.5.2 *Sample Processing*

The biofilm needs to be removed from the sampled medium within a few days of collection. The procedure used by Gray (1980) is summarized in Fig. 4.21. The sample bag is opened carefully, and the pieces of the medium are individually removed to prevent any of the animals, including individual flies, from escaping. The loose biofilm and associated animals are removed from the medium by gently brushing the surface with a soft artist's paintbrush in a shallow dish of water. At this stage, large animals present are removed, identified and counted, to prevent them from being damaged during the more vigorous washing procedure later on. The piece of medium is then placed into a beaker and stirred vigorously using a magnetic stirrer, and the next piece of medium is removed from the bag and initially brushed clean. The first piece of medium is then removed from the beaker and put into a second dish of water and scrubbed with a short-haired paintbrush (hogs hair, flat), which removes any tenacious biofilm still adhering to the surface. The medium and the sample bag are finally rinsed with distilled or deionized water to remove any remaining debris and the medium is then resealed in the bag. By standing the bags of freshly cleaned medium on a 30–40°C heater for about an hour, many animals, mainly enchytraeids and dipteran larvae, not removed from deep within the pores of pitted media such as slag or clinker, are driven to the surface by the heat. Both the bag and the medium are subsequently rewashed, and the extra animals collected are identified and counted. The bags of clean medium are finally left to dry completely and then reweighed. The weight of the biofilm is estimated by subtracting the dry weight of the bag and its contents from the wet weight, which had been measured immediately after the samples were returned to the laboratory.

All the debris, solids and animals removed from the sample of medium collected from each basket are added together and the total volume made up 500 $cm^3$ with distilled or deionized water. The liquid sample is then mixed thoroughly, and subsamples were taken for the various analyses summarized in Fig. 4.21. The subsamples taken for biological analysis are stored in glass bottles at 4°C. The remaining

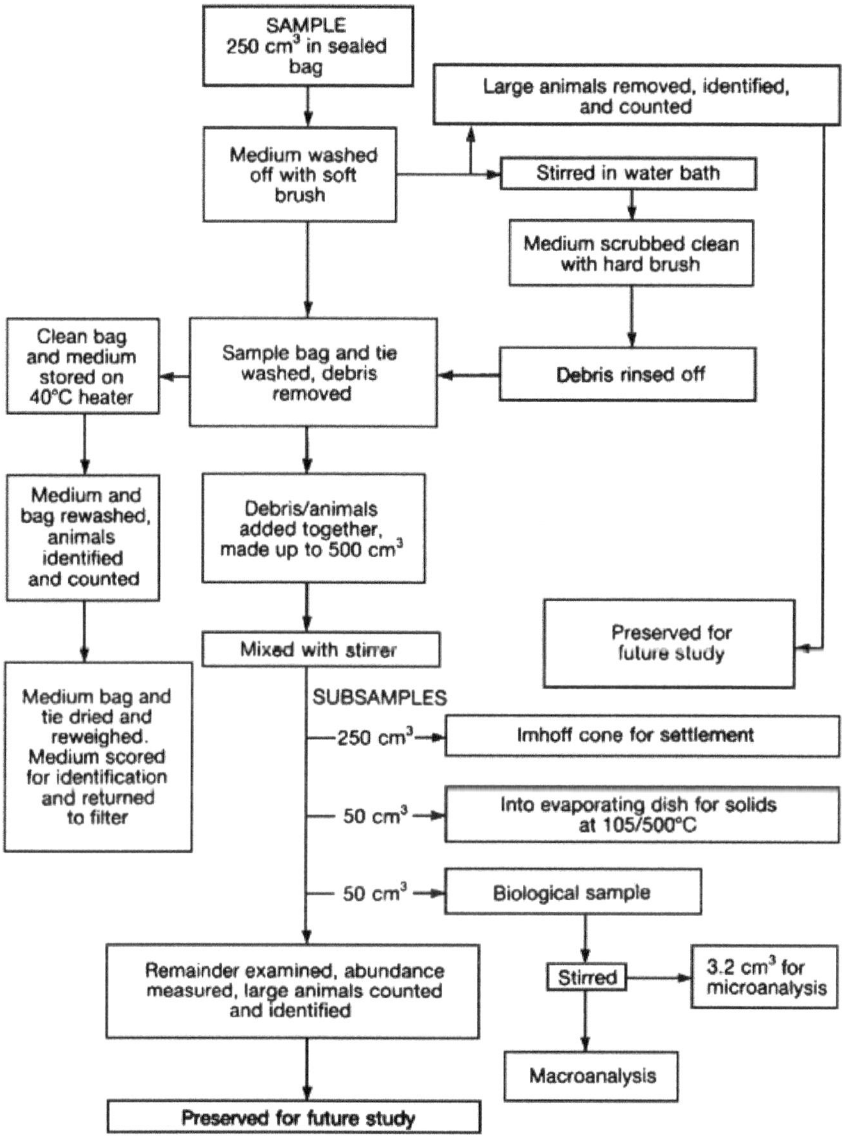


**Fig. 4.21.** Flow chart illustrating the processing of filter medium and associated film for chemical, physical and biological analysis.

150 $cm^3$, left after all the various subsamples are removed, is poured on to a white examination tray so that an assessment of the relative abundance of the various macroinvertebrates can be made. All the animals are identified, and the larger ones counted. Afterward, as much of the sample should be retained as possible and preserved for future study and reference.

An assessment of the volume of solids present can be made using the Imhoff cone method. The 250 $cm^3$ subsample (Fig. 4.21) is allowed to settle for 45 minutes in the cone, which is then gently twisted to remove any debris adhering to the glass sides. The quantity of solids settled after one hour are samples containing large numbers of organisms, and these should be recorded. Although most of the larger invertebrates should be removed before settlement, the samples contain large numbers of organisms, and these should be included in this assessment of biofilm accumulation.

### 4.5.3 *Micro and Meiofauna Analysis*

The 50 $cm^3$ subsample taken for biological analysis (Fig. 4.21) is shaken to produce complete mixing within the container. Then, using a sterile Pasteur pipette, a small volume is transferred to a haemacytometer type counting chamber of the modified Fuchs-Rosenthal type. A total area of 36 $mm^2$ split up into 0.25 $mm^2$ squares is examined for each sample under a compound microscope. Three magnifications are generally required: ×100 for counting large microorganisms such as large ciliate protozoa like *Paramecium caudatum*, Rotifera, nematodes and also large bacterial and fungal colonies, and ×200 for counting the other microorganisms that are normally identified at ×400. By measuring the depth of the sample under the cover slip of the counting chamber, the total volume of the sample examined per chamber can be calculated. Details of other counting chambers, staining methods, and separation methods for parasitic ova, cysts and nematodes are given by Fox *et al.* (1981) and the American Public Health Association *et al.* (2005).

The examination of bacteria comprising biofilms has been largely restricted to research and development rather than actual process management. Techniques developed primarily for activated sludge can also be applied to FFR biofilms. Phase contrast microscopy is still the first step in exploring the biofilm matrix for up to x400 magnification. Higher magnifications require the biomass to be fixed and usually stained. Filaments can be typed or occasionally identified to genera using the keys of Eikelboom (2006) and Jenkins *et al.* (2004), although further analysis requires more sophisticated microbial methods including confocal laser scanning microscopy (CLSM) and fluorescence *in situ* hybridization (FISH). The latter is now used with advanced staining techniques (e.g., DAPI and PHA staining) to help link specific species to function (Nielsen *et al.*, 2010). However, it is through molecular methods that bacterial identification is now principally carried out, generally using the 16S rRNA gene (Wojnowska-Baryła *et al.*, 2010). The use of DNA analysis in wastewater is reviewed by Karst *et al.* (2017), who also give detailed methodologies.

For accurate enumeration of the protozoa, Dubber and Gray (2009), working with activated sludge, recommend using a standard 25-μL subsample by using a standard slide with a 24 x 24 mm cover slip. They found that in order to have a 95% probability of recovering all the species, a very large number of replicates were required (i.e., 23–47). They recommended that by neglecting rare species, on the basis that they do not play an important role in the community structure, a smaller number of replicates could be used. Using a probability equation, the minimum number of replicates required can be determined after a pilot survey. They found six replicates gave excellent recovery (90–95%) of the species, excluding the three rarest, while analyzing just two replicates reduced the probability to just 25%, which rose to 50% for three replicates for the same recovery of the species.

The microfauna can only be identified accurately when alive, and this poses a practical problem regarding protozoans, in particular, because of their greater mobility. This is partly overcome by the addition of 1% of nickel sulfate to the sample, which has a narcotic effect on the protozoans. However, general use of this method should be avoided where possible as the peritrichs and the suctorians are more easily identified when active. Alternatively, small amounts of 2% xylocaine can be used to slow down the movement of protozoans without significantly affecting their physical form, thus allowing identification. The xylocaine will eventually evaporate and therefore treated samples should be processed reasonably quickly (Norouzian *et al.*, 1987). Microorganisms, including protozoa, cannot be stored for long periods or preserved for later analysis, so they must be examined as quickly as possible *in vivo*. Samples for microbial analysis should be gently aerated using a glass pasture pipette attached to an airline while waiting to be examined. This is best done in a small conical flask with cotton wool loosely packed around the pipette in the neck of the flask to prevent contamination. It is a best practice to identify and count protozoa as soon as possible (<3 hours) and certainly within 8 hours of collection as the community structure will change over time (Dubber and Gray, 2009).

Problems will arise in the identification of the fungi and filamentous bacteria, and in deciding how many cells or what length of filament constitutes the presence of a countable and reproducible unit. Gray (1983) sets minimum limits in his study. For a fungal hypha, this was 10 cells or 6 cells plus either a growing tip or conidium. For bacteria, only filaments longer than 0.2 mm or complete zoogloeal colonies were counted.

The main identification keys available for each major group are summarized in Table 4.10. Furthermore, Fox *et al.* (1981) have produced a beautifully illustrated key, which covers many groups, including parasitic helminths, parasitic and free-living protozoa, algae, and many other meiofauna groups. However, this is not a key in the true sense but rather a collection of photomicrographs of microorganisms isolated at the sewage treatment plants of the Metropolitan Sanitary District of Greater Chicago. Digital photographs can be a useful aid for identification, and it can prove invaluable for monitoring the changes in the surface biofilm. A comprehensive selection of wastewater analysis, designed for activated sludge but also applicable to the study of FFRs, such as the measurement of extracellular polymeric substances (EPS), has been collated by Loosdrecht *et al.* (2017). An example of biofilm characterization is given by Andersson (2009).

**Table 4.10.** The main identification keys for each major biological group found in fixed-film reactors.

| Microfauna group | Key references |
|---|---|
| Bacteria | Eikelboom, 2006; Eikelboom and Van Buijsen, 1981; Jenkins *et al.*, 2004 |
| Fungi | Cooke, 1963; Tomlinson and Williams, 1975 |
| Algae | Belcher and Swale, 1976; George, 1976; Bellinger, 1980; Sykes, 1981 |
| Protozoa | Martin, 1968; Calaway and Lackey, 1962; Bick, 1972; Page, 1976; Curds, 1969; 1982; Curds *et al.*, 1983; Foissner and Berger, 1996 |
| Nematoda | Tarjan *et al.*, 1977 |
| Rotifera | Donner, 1966; Ruttner-Kolisko, 1972; Pontin, 1978 |
| Annelida | Brinkhurst, 1971; Nielson and Christensen, 1959; 1961; 1963; Gerard, 1964; Sperber, 1950; Tynen, 1966 |
| Insecta | Lawrence, 1970; Satchell, 1947; 1949; Coe *et al.*, 1950; Bryce, 1960; Brindle, 1962; Mason, 1968; Bryce and Hobart, 1972; Unwin, 1981 |
| Arachnida | Evans *et al.*, 1961 |
| Crustacea | Harding and Smith, 1974 |
| Mollusca | Janus, 1965; Cameron *et al.*, 1983 |

*General reference work:* Armitage *et al.*, 1979; Martin, 1968; Tomlinson, 1946a.

## 4.5.4 *Macrofauna Analysis*

The remainder of the subsample used for the microfaunal analysis is shaken as before and poured into a large Hartley pattern Buchner funnel, and the sample container rinsed out. The sample is gently filtered at low pressure so as not to damage the annelids, through Whatman 113 (150 mm diameter) filter paper. The filter paper is then cut into four equal sections and examined in a low form plastic dish under a stereo microscope. The larger invertebrates, such as dipteran larvae and enchytraeid worms can be identified, counted and removed at ×5–10 magnification, whereas the other invertebrates have to be located before identification and counted by a systematic search using fine needles at ×15–20 magnification. Mites must be identified and counted separately using a 1 $cm^2$ illuminated background plate that fits under the plastic dish containing the filter paper. This allows only a specific area to be illuminated, which can then be carefully searched at ×40 magnification. A total area of 6 cm$^2$ is searched in this way, the 1-cm$^2$ areas being chosen at random on the four sections of the filter paper. Apart from the general key by Tomlinson (1946a), which covers only a few of the most common percolating filter grazing organisms and is now taxonomically very out of date, there is no general key available. Therefore, identification must be made using specialized keys for each major group. The number of grazers is expressed as the total number per liter or cubic meter of the medium.

Curds and Cockburn (1970) found that a greater variety of protozoan species were to be found in the effluent from filters than in the biofilm collected just from the surface of the filter. Similarly, Learner (1975a) recorded only 39 invertebrate grazing species in surface media samples (0–30 cm depth) and a further 48 in the final effluent. Therefore, to obtain a comprehensive list of species present in the filter and to discover which micro and macroorganisms are being washed out of the filters, the effluent should be regularly examined. The same methods are used for the effluent and the media. Unfiltered effluent can be collected as spot 250 mL samples for protozoan analysis or larger species by passing 5 or 10 liters of effluent through a drift net with a 400-μm mesh and collection bottle attached to the end.

To assess the rate at which flies emerge from a filter a special emergence trap is required (Solbe *et al.*, 1967) (Fig. 4.17). Styles (1979) developed a much simpler trap of smaller dimensions, which is ideal for trapping Chironomidae. It is a rectangular perspex box that is 280 × 140 mm and 70 mm high with a removable sliding lid that is covered with fly adhesive on the under surface, which acts as the trapping area. Each end of the box is made of nylon mesh to allow ventilation. It was designed specially to be placed in the inter-jet zone between the jets from the distributor and does not incorporate a system of redistributing the wastewater, as seen in the larger traps. An accurate assessment of the aerial density of flies over or near the filter can be more easily obtained. A variety of traps have been designed and successfully used (Tomlinson and Stride, 1945; Hawkes, 1951; Taylor, 1951; 1955; Hawkes, 1983). Although traps are by far the most accurate and efficient method of measuring aerial density, they are expensive to construct. However, large sticky paper sheets stretched out between supports are simple and fairly accurate, although identification of the smaller species can be difficult because the flies, as they struggle, become covered in the adhesive used on the paper. Fly abundance can also be assessed by using a large entomological aspirator to catch as many flies as possible over a standard collection period.

## References

American Public Health Association, American Water Works Association and Water Pollution Control Federation (2005). *Standard Methods for the Examination of Water and Wastewater, 21st ed.* American Public Health Association Inc., New York.

Andersson, S. (2009) Characterization of bacterial biofilms for wastewater treatment. Royal Institute of Technology, School of Biotechnology, Stockholm, Sweden. https://www.diva-portal.org/smash/get/diva2:209486/FULLTEXT01.pdf

Armitage, P.D., Furse, M.T. and Wright, J.F. (1979) A bibliography of works for the identification of freshwater invertebrates in the British Isles. Occasional Publication 5. Freshwater Biological Association, Cumbria.

Baines, S., Hawkes, H.A., Hewitt, C.H. and Jenkins, S.H. (1953) Protozoa as indicators in activated sludge treatment. *Sewage and Industrial Wastes*, **25**, 1023–1033.

Baker, M.N. (1949) *The Quest for Pure Water.* American Water Works Association, New York.

Baker, R.A. (1961) A preliminary survey of the mite fauna of sewage percolating filters. M.Sc. Thesis. University of London.

Baker, R.A. (1975) Arachnida. In *Ecological Aspects of Used-Water Treatment* (eds. C.R. Curds and H.A. Hakes). Academic Press, London, pp. 375–392.

Barker, A.N. (1942) The seasonal incidence, occurrence and distribution of protozoa in the bacteria bed process of sewage disposal. *Annals of Applied Biology*, **29**, 23–33.

Barker, A.N. (1946) The ecology and function of protozoa in sewage purification. *Annals of Applied Biology*, **33**, 314–325.

Barritt, N.W. (1940) The ecology of activated sludge in relation to its properties and the isolation of a specific substance from the purified effluent. *Annals of Applied Biology*, **27**, 151–156.

Becker, J.H. and Shaw, C.G. (1955) Fungi in domestic sewage treatment plants. *Applied Microbiology*, **3**, 173–180.

Belcher, H. and Swale, E. (1976) *A Beginners Guide To Freshwater Algae.* Institute of Terrestrial Ecology, H.M.S.O., London.

Bellinger, E.G. (1980) *A Key To Common British Algae.* Institution of Water Engineers and Scientists, London.

Benson-Evans, K. and Williams, P.F. (1975) Algae and bryophytes. In *Ecological Aspects of Used-Water Treatment. Vol. I. The Organisms and Their Ecology* (eds. C.R. Curds and H.A. Hawkes). Academic Press, London, pp. 153–202.

Bergtold, M., Mayr, G. and Traunspurger, W. (2007) Nematodes in wastewater biofilms — Appearance and density of species in three biofilter reactors. *Water Research,* **41**, (1), 145–151.

Bick, H. (1972) *Ciliated Protozoa: An Illustrated Guide to the Species used as Biological Indicators in Freshwater Biology.* World Health Organization, Geneva.

Brindle, A. (1962) Taxonomic notes on the larvae of British Diptera: 2. Trichoceridae and Anisopodidae. *Entomologist*, **95**, 285–288.

Brink, N. (1967) Ecological studies in biological filters. *Internationale Revue der Gesamten Hydrobiologie*, **52**, (1), 51–122.

Brinkhurst, R.O. (1971) *A Guide for the Identification of British Aquatic Oligochaeta.* Scientific Publication, 22. Freshwater Biological Association, Cumbria, UK.

Bruce, A.M. and Merkens, J.C. (1970) Recent studies of high-rate biological filtration. *Water Pollution Control*, **69**, 113–148.

Bruce, A.M., Merkens, J.C. and Haynes, B.A.O. (1975) Pilot-scale studies on the treatment of domestic sewage by two-stage biological filtration with special reference to nitrification. *Water Pollution Control*, **74**, 80–100.

Bruce, A.M., Merkens, J.C. and MacMillan, S.C. (1970) Research development in high-rate biological filtration. *Institution of Public Health Engineers Journal*, **69**, 178–207.

Bryce, D. (1960) Studies on the larvae of the British Chironomidae (Diptera), with keys to the Chironomidae and Tanypodire. *Transactions of the Society for British Entomology*, **14**, (II), 19–61.

Bryce, D. and Hobart, A. (1972) The biology and identification of the larvae of the Chironomidae (Diptera). *Entomologists Gazette*, **23**, 175–217.

Bungay, H.R. and Bungay, M.L. (1968) Microbial interactions in continuous culture. In *Advances in Applied Microbiology, Vol. 10* (eds. W.W. Umbreit and D. Perlman). Academic Press, London, pp. 269–290.

Calaway, W.T. (1968) The metazoa of waste treatment processes: Rotifers. *Journal of the Water Pollution Control Federation*, **40**, 412–422.

Calaway, W.T. and Lackey, J.B. (1962) *Waste Treatment Protozoa — Flagellata*. University of Florida Engineering Series 3.

Cameron, R.A.D., Jackson, N. and Eversham, B. (1983) A field key to the slugs of the British Isles. *Field Studies*, **5**, 807–824.

Chaudhuri, N., Engelbrecht, R.S. and Austin, J.H. (1965) Nematodes in an aerobic waste treatment plant. *Journal of the American Water Works Association*, **57**, 1561.

CIWEM (2000) *Biological Filtration and Other Fixed Film Processes*. The Chartered Institution of Water and Environmental Management, London, UK.

Clay, E. (1964) The fauna and flora of sewage processes: 2. Species list. Memorandum PVM 45//732, Research Department, Paints Division, ICI Ltd.

Coe, R.L., Freeman, P. and Mattingley, P.E. (1950) *Diptera: Nematorea. Identification Handbook of British Insects*, **9**, (2). British Entomological Society.

Cooke, W.B. (1954) Fungi in polluted water and sewage. II. Isolation technique. *Sewage and Industrial Wastes*, **26**, 661–674.

Cooke, W.B. (1959) Trickling filter ecology. *Ecology*, **40**, 273–291.

Cooke, W.B. (1963) *A Laboratory Guide to Fungi in Polluted Waters, Sewage and Sewage Treatment Works*. Public Health Service Publication No. 999-WP-1, US Department of Health, Education and Welfare.

Coombs, R.M., Cleworth, M.A. and Davies, D.H. (1996) The control of *Psychoda alternate* (Psychodidae) in sewage biological filters by application of the insect growth regulator Pyriproxyfen. *Water Research*, **30**, 654–662.

Coombs, R.M., Dancer, B.N., Davies, D.H., Houston, J. and Learner, M.A. (1991) The use of *Bacillus thuringiensis var. israelensis* to control the nuisance fly *Sylvicola fenestralis* (Anisopodidae) in sewage filter beds. *Water Research*, **25**, 605–611.

Curds, C.R. (1969) *An Illustrated Key to the British Freshwater Ciliated Protozoa Commonly Found in Activated Sludge*. Water Pollution Research Technical Paper No. 12. Water Pollution Research Laboratory, Ministry of Technology. HMSO, London.

Curds, C.R. (1973) The role of protozoa in the activated sludge process. *American Zoology*, **13**, 161–169.

Curds, C.R. (1975) Protozoa. In *Ecological Aspects of Used-Water Treatment, Vol. 1. The Organisms and their Ecology* (eds. C.R. Curds and H.A. Hawkes). Academic Press, London, pp. 203–268.

Curds, C.R. (1982) British and other freshwater ciliated protozoa: Part I. In *Synopses of the British Fauna*, 22. Linnean Society of London, Cambridge University Press, Cambridge.

Curds, C.R. and Cockburn, A. (1970) Protozoa in biological sewage-treatment processes: I. A survey of the protozoan fauna of British percolating filters and activated sludge plants. *Water Research*, **4**, 225–236.

Curds, C.R. and Hawkes, H.A. (eds.) (1975) *Ecological Aspects of Used-Water Treatment, Vol. 1. The Organisms and their Ecology*. Academic Press, London.

Curds, C.R. and Vandyke, J.M. (1966) The feeding habits and growth rates of some freshwater ciliates found in activated sludge plants. *Journal of Applied Ecology*, **3**, 127–137.

Curds, C.R., Gates, M.A., and McL. Roberts, D. (1983) British and other freshwater ciliated protozoa: Part II. In *Synopses of the British Fauna*, **23**, Linnean Society of London, Cambridge University Press, Cambridge.

Cutler, D.W., Crump, L.M. and Dixon, A. (1932) Some factors influencing the distribution of certain protozoa in biological filters. *Journal of Animal Ecology*, **1**, 141–151.

Donner, J. (1966) Rotifers (trans. H.G.S. Wright). Frederick Warne, London.

Doohan, M. (1975) Rotifera. In *Ecological Aspects of Used Water Treatment, Vol. 1. The Organisms and their Ecology* (eds. C.R. Curds and H.A. Hawkes). Academic Press, London, pp. 289–304.

Dubber, D. and Gray, N.F. (2009) Enumeration of protozoan ciliates in activated sludge: Determination of replicate number using probability. *Water Research*, **43**, 3443–3452.

Duffey, E. (1997) Spider adaptations to artificial biotopes: The fauna of percolating filter beds in a sewage treatment works. *Journal of Applied Ecology*, **35**, (5), 1190–1202.

Dyson, J.E.B. and Lloyd, L. (1936) The distribution of the early stages of *Metriocnemus longitarus* Goet, (Chironomidae), in sewage bacteria beds. *Proceedings of the Leeds Philosophical and Literature Society*, **3**, 174–176.

Eikelboom, D.H. (2006) *Identification and Control of Filamentous Microorganisms in Industrial Wastewater Treatment Plants.* IWA Publishing, London, UK.

Eikelboom, D.H. and Van Buijsen, H.J.J. (1981) Microscopic sludge investigation manual. TNO Research Institute for Environmental Hygiene, Delft.

Evans, G.O., Sheals, J.G. and MacFarlane, D. (1961) *The Terrestrial Acari of the British Isles*, **1**. British Museum of Natural History, London.

Foissner, W. and Berger, H. (1996) A user friendly guide to the ciliates (Protozoa, Ciliophora) commonly used by hydrobiologists as bioindicators in rivers, lakes, and wastewaters, with notes on their ecology. *Freshwater Biology*, **35**, 375–482.

Fox, J.C., Fitzgerald, P.R. and Lue-Hing, C. (1981) Sewage organisms: A color atlas. The Metropolitan Sanitary District of Greater Chicago, Chicago, US.

Frye, W.W. and Becker, E.R. (1929) The fauna of an experimental trickling filter. *Sewage Works Journal*, **1**, 286–308.

George, E.A. (1976) A guide to algal keys (excluding seaweeds). *British Phycological Journal*, **11**, 49–55.

Gerard, B.M. (1964) *Synopses of British Fauna, Vol. 6. Lumbricidae (Annelida)*. The Linnean Society of London.

Gray, N.F. (1980) The comparative ecology of high-rate plastic, conventional mineral and mixed plastic/mineral media in the treatment of domestic sewage in percolating filters. Vol. I. Ph.D. Thesis. Council for National Academic Awards.

Gray, N.F. (1982) The use of percolating filters in teaching ecology. *Journal of Biological Education*, **16**, 183–186.

Gray, N.F. (1983) Ponding of a random plastic filter medium due to the fungus *Subbaromyces splendens* Hesseltine in the treatment of sewage. *Water Research*, **17**, 1295–1302.

Gray, N.F. (1984) Ecological consequences of *Ambystegium riparium* growing on a sewage work's percolating filter. *Bulletin of the Irish Biogeographical Society*, **8**, 26–30.

Gray, N.F. and Learner, M.A. (1983) A pilot-scale percolating filter for use in sewage treatment studies. *Water Research*, **17**, 249–253.

Harding, J.P. and Smith, W.A. (1974) *A Key to the British Freshwater Cyclopid and Calanoid Copepods*. Scientific Publications 18, Freshwater Biological Association, Cumbria.

Harkness, N. (1966) Bacteria in sewage treatment processes. *Journal and Proceedings of the Institute of Sewage Purification*, pp. 542–557.

Hawkes, H.A. (1951) A study of the biology and control of *Anisopus fenestralis* (Scopoli, 1763), a fly associated with sewage filters. *Annals of Applied Biology*, **38**, 592–605.

Hawkes, H.A. (1952) The ecology of *Anisopus fenestralis* Scop. (Diptera) in sewage bacteria beds. *Annals of Applied Biology*, **39**, 181–192.

Hawkes, H.A. (1955) The effect of periodicity of dosing on the amount of film and the numbers of insects and worms in alternating filters at Minworth. *Journal and Proceedings of the Institute of Sewage Purification*, (1), 48–58.

Hawkes, H.A. (1957) Film accumulation and grazing activity in the sewage filters at Birmingham. *Journal and Proceedings of the Institute Sewage Purification*, (1), 88–110.

Hawkes, H.A. (1959) The effects of methods of sewage application on the ecology of bacteria beds. *Annals of Applied Biology*, **37**, 339–349.

Hawkes, H.A. (1961) Fluctuations in the aerial density of *Anisopus fenestralis* Scop. (Diptera) above sewage bacterial beds. *Annals of Applied Biology*, **49**, 66–76.

Hawkes, H.A. (1963) *The Ecology of Wastewater Treatment*. Pergamon, Oxford.

Hawkes, H.A. (1965) Factors influencing the seasonal incidence of fungal growths in sewage bacteria beds. *International Journal of Air and Water Pollution*, **9**, 693–714.

Hawkes, H.A. (1983) The applied significance of ecological studies of aerobic processes. In *Ecological Aspects of Used Water Treatment, Vol. 3. The Processes and their Ecology* (eds. C.R. Curds and H.A. Hawkes). Academic Press, London, pp. 173–334.

Hawkes, H.A. and Jenkins, S.H. (1951) Biological principles in sewage purification. *Journal and Proceedings of the Institute of Sewage Purification*, (3), 300–318.

Hawkes, H.A. and Jenkins, S.H. (1955) Comparison of four grades of sewage percolating filter media in relation to purification, film accumulation and fauna. *Journal and Proceedings of the Institute of Sewage Purification*, (3), 352–357.

Hawkes, H.A. and Jenkins, S.H. (1958) Comparison of four grades of filter media in relation to purification, film accumulation and fauna of sewage percolating filters operating on alternating double filtration. *Journal and Proceedings of the Institute Sewage Purification*, (2), 221–225.

Hawkes, H.A. and Shepherd, M.R.N. (1972) The effect of dosing frequency on the seasonal fluctuations and vertical distribution of solids and grazing fauna in sewage percolating filters. *Water Research*, **6**, 721–730

Hemming, M.L. and Wheatley, A.D. (1979) Low rate biofiltration systems using random plastic media. *Water Pollution Control*, **78**, (1), 54–68.

Hesseltine, C.W. (1953) Study of trickling filter fungi. *Bulletin of the Torrey Botanical Club*, **80**, 507–514.
Heukelekian, H. (1947) Use of direct method of oxygen utilization in waste treatment studies. *Sewage Works Journal*, **19**, 875–882.
HMSO (1988) Examining biological filters, toxocity to aerobic bacteria, effect of SRT and temperature 1985–6. Methods for the Examination of Waters and Associated Materials, Her Majesty's Stationery Office London.
Holtje, R.H. (1943) The biology of sewage sprinkling filters. *Sewage Works Journal*, **15**, 14–29.
Houston, J., Dancer, B.N. and Learner, M.A. (1989a) Control of sewage filter flies using *Bacillus thuringiensis var. israelensis*-I. Acute toxicity tests and pilot scale trial. *Water Research*, **23**, 369–378.
Houston, J., Dancer, B.N. and Learner, M.A. (1989b) Control of sewage filter flies using *Bacillus thuringiensis var. israelensis*-II. Full scale trials. *Water Research*, **23**, 379–385.
Hughes, A.M. (1961) *The Mites of Stored Food Products.* HMSO, London.
Hussey, B.R. (1975) Ecological studies on percolating filters and stream riffles associated with the disposal of domestic and industrial wastes. M.Sc. Thesis. University of Aston in Birmingham, UK.
Hussey, B.R. (1982) Moss growth in filter beds. *Water Research*, **16**, 391–398.
Ingram, W.T. and Edwards, G.P. (1960) The behaviour of filter biota under controlled conditions. *Proceedings of the Third Biological Waste Treatment Conference.* Manhattan College, New York.
IWPC (1988) *Biological Filtration.* Institute of Water Pollution Control, Maidstone, UK.
James, A. (1964) The bacteriology of trickling filters. *Journal of Applied Bacteriology*, **27**, (2), 197–207.
Janus, H. (1965) *The Young Specialist Looks at Land and Freshwater Molluscs.* Burke, London.
Jenkins, D., Richard, M.G., and Daigger, G.T. (2004) *Manual on the Causes and Control of Activated Sludge Bulking, Foaming, and Other Solids Separation Problems, 3rd ed.* IWA Publishing, London, UK.
Jiang, J.Q., Stabedakis, K.S., Thomas, D.N. and Dussart, G.B.J. (2003) A trial on the use of biological aerated filter (BAF) technology for combating the "snail problem". In *Slugs and Snails: Agricultural, Veterinary and Environmental Perspectives* (ed. G.B.J. Dussart). *British Crop Protection Council Symposium Proceedings*, pp. 249–254.
Jones, M.B. and Wigham, G.D. (1993) Reproductive-biology of *Orchestia-gammarellus* (Crustacea, Amphipodia) living in a sewage-treatment

works. *Journal of the Marine Biological Association of the United Kingdom*, **73**, (2), 405–416.

Joslin, J.R., Sidwick, J.M., Greene, C. and Shearer, J.R. (1971) High rate biological filtration: A comparative assessment. *Water Pollution Control*, **70**, 383–399.

Kamei, M., Kamada, A., Utsumi, Y. and Ishi, T. (1993) Laboratory and field evaluation of methoprene and its slow release formulation, Altosid 10F, against the sprinkler sewage fly *Psychoda alternata* Say (Diptera: Psychodidae). *Applied Entomological Zoology*, **28**, 19–25.

Karst, S.M., Albertsen, M., Kirkegaard, R.H., Dueholm, M.S. and Nielsen P.H. (2017) Molecular methods. In *Experimental Methods in Wastewater Treatment* (eds. M.C.M. van Loosdrecht, P.H. Nielsen, C.M. Lopez-Vazquez and D. Brdjanovic). IWA Publishing, London, UK.

Kocerba-Soroka, W., Fiałkowska, E., Pajdak-Stós, A., Klimek, B., Kowalska, E., Drzewicki, A., Salvadó, H. and Fyda, J. (2013) The use of rotifers for limiting filamentous bacteria Type 021N, a bacteria causing activated sludge bulking. *Water Science and Technology*, **67**, (7), 1557–1563.

Kowalska, E., Paturej, E. and Zielińska, M. (2014) Use of *Lecane* rotifers for limiting *Thiothrix* filamentous bacteria in bulking activated sludge in a dairy wastewater treatment plant. *Archives of Biological Sciences*, **66**, (4), 1371–1378.

Kowalska, E., Paturej, E. and Zielińska, M. (2016) Use of *Lecane inermis* for control of sludge bulking caused by the *Haliscomenobacter* genus. *Desalination and Water Treatment*, **57**, (23), 10916–10923.

Lackey, J.B. (1924) Studies of the fauna of Imhoff tanks and sprinkling beds. *Bulletin of the New Jersey Agricultural Experimental Station*, **403**, 40–60.

Lackey, J.B. (1925) The fauna of Imhoff tanks. I. Ecology of Imhoff tanks. *Bulletin of the New Jersey Agricultural Experimental Station*, **417**, 1–39.

Lawrence, P.N. (1970) Collembola (Spring tails) of sewage filters. *Water and Waste Treatment*, **13**, 106–109.

Learner, M.A. (1972) Laboratory studies on the life-histories of four enchytraeid worms (Oligochaeta) which inhabit sewage percolating filters. *Annals of Applied Biology*, **70**, 251–266.

Learner, M.A. (1975a) The ecology and distribution of invertebrates which inhabit the percolating filters of sewage-works. Ph.D. Thesis. University of London.

Learner, M.A. (1975b) Insecta. In *Ecological Aspects of Used-Water Treatment* (eds. C.R. Curds and H.A. Hawkes). Academic Press, London, pp. 337–374.

Learner, M.A. (2000) Egression of flies from sewage filter-beds. *Water Research*, **34**, (3), 877–889.

Learner, M.A. and Chawner, H.A. (1998) Macro-invertebrate associations in sewage filter-beds and their relationship to operational practice. *Journal of Applied Ecology*, **35**, 720–747.

Liang, P., Huang, X. and Qian, Y. (2006) Excess sludge reduction in activated sludge through predation of *Aeolosoma hemprichi*. *Biochemical Engineering Journal*, **28**, 117–122.

Liebmann, H. (1949) The biology of percolating filters. *Vom Wasser*, **17**, 62–82.

Liebmann, H. (1951) *Handbuch der Frishwasser-und Abwasser-biologie*. Gustav, Fischer, Jena.

Lloyd, L. (1945) Animal life in sewage purification processes. *Journal and Proceedings of the Institute of Sewage Purification*, pp. 119–139.

Lloyd, L., Graham, J.F. and Reynoldson, T.B. (1940) Materials for a study in animal competition. The fauna of the sewage bacteria beds. *Annals of Applied Biology*, **27**, 122–150.

Loosdrecht, M.C.M. van, Nielsen, P.H., Lopez-Vazquez, C.M. and Brdjanovic, D. (2017) *Experimental Methods in Wastewater Treatment*. IWA Publishing, London, UK.

Lumb, B.C. and Eastwood, P.K. (1958) The recirculation principle in filtration of settled sewage — Some comments on its application. *Journal and Proceedings of the Institute Sewage Purification*, pp. 380–398.

Martin, R. (1968) Microfauna of biological filters. *Civil Engineering Bulletin*, **39**.

Mason, W.T. (1968) *An Introduction to the Identification of Chironomid Larvae*. Division of Pollution Surveillance, Federal Water Pollution Control Administration, US Department of the Interior.

Mistri, M., Gavinelli, M. and Gorni, A. (1994) Structure of a protozoan community in a percolating filter. *Bollettion Di Zoologia*, **61**, (1), 65–71.

Moser, H. (1958) *The Dynamics of Bacterial Population Maintained in the Chemostat*. Carnegie Institute, Publication 614. Washington, DC.

Murad, J.L. and Bazer, G.T. (1970) Diplogasterid and rhabditid nematodes in a waste-water treatment plant and factors related to their disposal. *Journal of the Water Pollution Control Federation*, **42**, (1), 105–114.

Nielsen, C.O. and Christensen, B. (1959) The Enchytraeidae: Critical revision and taxonomy of European species. *Natura Jutlandica*, **8/9**, 1–160.

Nielsen, C.O. and Christensen, B. (1961) The Enchytraeidae: Critical revision and taxonomy of European species. Suppl. I. *Natura Jutlandica*, **10**, 1–23.

Nielsen, C.O. and Christensen, B. (1963) The Enchytraeidae: Critical revision and taxonomy of European species. Suppl. II. *Natura Jutlandica*, **3**, 1–20.

Nielsen, J.L., Kragelund, C. and Nielsen, P.H. (2010) Ecophysiological analysis of microorganisms in complex microbial systems by combination of fluorescence *in situ* hybridization with extracellular staining techniques. *Methods in Molecular Biology*, **599**, 117–128.

Norouzian, M., Herroz-Zamorano, A. and Perea-Mejia, C. (1987) A technique for the enumeration of protozoa in wastewater. *Environmental Technology Letters*, **8**, 221–224.

Otter, C.S. den (1966) A physical method for permanent control of *Psychoda* pests at wastewater treatment plants. *Journal of the Water Pollution Control Federation*, **38**, 156–164.

Page, F.C. (1976) An illustrated key to freshwater and soil amoebae. Scientific Publication 34. The Freshwater Biological Association, Cumbria.

Painter, H.A. (1970) A review of the literature on inorganic nitrogen metabolism. *Water Research*, **4**, 393–450.

Painter, H.A. (1980) A survey of filter fly nuisances and their remedies. Technical Report TR155. Water Research Centre, Stevenage.

Pajdak-Stós, A., Kocerba-Soroka, W. and Fyda, J. (2017) Foam-forming bacteria in activated sludge effectively reduced by rotifers in laboratory- and real-scale wastewater treatment plant experiments. *Environmental Science and Pollution Research*, **24**, 13004–13011.

Palfrey, E., Coombs, R. and Davies, D. (1992) New methods of fly control for percolating filters. WRc Report: UM1294. Water Research Centre Publications, Swindon, UK.

Peters, B.G. (1930) Some nematodes met with in a biological investigation of sewage. *Journal of Helminthology*, **8**, 165–184.

Petersen, J.J. (1982) Current status of nematodes for biological control of insects. *Parasitology*, **84**, 177–204.

Pike, E.B. (1975) Aerobic bacteria. In *Ecological Aspects of Used-Water Treatment, Vol. 1. The Organisms and their Ecology* (eds. C.R. Curds and H.A. Hawkes). Academic Press, London, pp. 1–64.

Pike, E.B. and Carrington, E.G. (1972) Recent developments in the study of bacteria in the activated-sludge process. *Water Pollution Control*, **71**, 583–605.

Pillai, J.K. and Taylor, D.P. (1968) *Butlerlius micans* N. sp. (Nematoda: Diplogasterinae) from Illinois, with observations on its feeding habits and a key to the species of Butlerius Goodey, 1929. *Nematologica*, **14**, 89–93.

Pontin, R.M. (1978) *A Key to the Freshwater Planktonic and Semi-Planktonic Rotifera of the British Isles.* Scientific Publications, 38. Freshwater Biological Association, Cumbria, UK.

Reynoldson, T.B. (1941) The biology of the macrofauna of a high rate double filtration plant at Huddersfield. *Journal and Proceedings of the Institute of Sewage Purification*, pp. 109–128.

Reynoldson, T.B. (1947) An ecological study of the enchytraeid wormpopulation of sewage bacteria beds: Feld investigations. *Journal of Animal Ecology*, **16**, 26–37.

Reynoldson, T.B. (1948) An ecological study of the enchytraeid worm population of sewage bacteria beds: Synthesis of field and laboratory data. *Journal of Animal Ecology*, **17**, 27–38.

Rowlands, C.L. (1979) Ecological and comparative performance studies on high-rate filter media. Ph.D. Thesis. University of Aston in Birmingham, UK.

Ruttner-Kolisko, A. von (1972) *Rotatoria.* In *Die Binnengewässer*, 26, Das Zooplankton der Binnengewässer. 1, Teil., pp. 99–225. E. Schweizerbart'sche Verlagsbuch-handlung, Stuttgart.

Satchell, G.H. (1947) The larvae of the British species of *Psychoda* (Diptea: Psychodidae). *Parasitology*, **38**, 57–69.

Satchell, G.H. (1949) The respiratory horns of *Psychoda* pupae (Diptera: Psychodidae). *Parasitology*, **39**, 43–52.

Scherb, K. (1968) Nematoda. In *Troptkorper und Belebungsbecken* (ed. H. Liebmann). R. Oldenbourg, Munich, pp. 158–206.

Schiemer, F. (1975) Nematoda. In *Ecological Aspects of Used-Water Treatment, Vol. 1. The Organisms and their Ecology* (eds. C.R. Curds and H.A. Hawkes). Academic Press, London, pp. 269–288.

Sladká, A. and Ottova, V. (1968) The most common fungi in biological treatment plants. *Hydrobiologia*, **31**, 350–362.

Solbe, J.F. de L.G. (1971) Aspects of the biology of the lumbricids *Eiseniella tetraedra* (Savigny) and *Dendrobaena rubida* (Savigny) and *F. subarubicunda* (Eisen) in a percolating filter. *Journal of Applied Ecology*, **8**, 845–867.

Solbe, J.F. de L.G. (1975) Annelida. In *Ecological Aspects of Used-Water Treatment, Vol. 1. The Organisms and their Ecology* (eds. C.R. Curds and H.A. Hawkes). Academic Press, London, pp. 305–355.

Solbe, J.F. de L.G. and Tozer, J.S. (1971) Aspects of the biology of *Psychoda alternata* (Say.) and *P. severini parthenogenetica* Tonn. (Diptera) in a percolating filter. *Journal of Applied Ecology*, **8**, 835–844.

Solbe, J.F. de L.G., Ripley, P.G. and Tomlinson, T.G. (1974) The effect of temperature on performance of experimental percolating filters with and without mixed macro-invertebrate populations. *Water Research*, **8**, 557–573.

Solbe, J.F. de L.G., Williams, N.V. and Roberts, H. (1967) The colonization of a percolating filter by invertebrates, and their effect on settlement of humus solids. *Water Pollution Control*, **66**, (5), 423–448.

Sperber, C. (1950) A guide for the determination of European Naididae. *Zoologiste Bidrag Fran Uppsala*, **29**, 45–78.

Styles, P.D. (1979) Control of chironomid flies breeding in sewage filters. Ph.D. Thesis, University of Aston in Birmingham.

Sydenham, D.H.J. (1968) The Ecology of Protozoa and Other Organisms in Activated Sludge. Ph.D. Thesis. University of London.

Sydenham, D.H.J. (1971) A re-assessment of the relative importance of ciliates, rhizopods, and rotalorians in the ecology of activated sludge. *Hydrobiologia*, **38**, 553–563.

Sykes, J.B. (1981) An illustrated guide to the Diatoms of British coastal plankton. *Field Studies*, **5**, 425–468.

Tarjan, A.C., Esser, R.P. and Chang, S.C. (1977) An illustrated key to nematodes found in freshwater. *Journal of the Water Pollution Control Federation*, **49**, 2318–2337.

Taylor, L.R. (1951) An improved suction trap for insects. *Annals of Applied Biology*, **38**, 582–591.

Taylor, L.R. (1955) The standardization of air-flow in insect suction traps. *Annals of Applied Biology*, **43**, 390–408.

Terry, R.J. (1951) The behaviour and distribution of the larger worms in trickling filters. *Journal and Proceedings of the Institute of Sewage Purification*, pp. 16–25.

Terry, R.J. (1956) The relations between bed medium and sewage filters and the flies breeding in them. *Journal of Animal Ecology*, **25**, 6–14.

Thomas, D. and Dussart, G.B.J. (2003) Sewage works failure because of invasive snails — Problem and solution. In *Slugs and Snails: Agricultural,*

*Veterinary and Environmental Perspectives* (ed. G.B.J. Dussart). *British Crop Protection Council Symposium Proceedings*, pp. 109–114.

Tomlinson, T.G. (1942) The nitrifying activity of biological film from different depths in percolating filters treating sewage by the processes of alternating double filtration and single filtration. *Proceedings of the Society of Agricultural Bacteriologists*, pp. 5–7.

Tomlinson, T.G. (1946a) Animal life in percolating filters. Identification of flies, worms and some other common organisms. Water Pollution Research Technical Paper 9. Department of Scientific and Industrial Research, HMSO, London.

Tomlinson, T.G. (1946b) The growth and distribution of film in percolating filters treating sewage by single and alternating double filtration. *Journal and Proceedings of the Institute of Sewage Purification*, pp. 168–183.

Tomlinson, T.G. and Hall, H. (1950) Some factors in the treatment of sewage in percolating filters. *Journal and Proceedings of the Institute of Sewage Purification.*, pp. 338–355.

Tomlinson, T.G. and Snaddon, D.H.M. (1966) Biological oxidation of sewage by films of micro-organisms. *International Journal of Air and Water Pollution*, **10**, 865–881.

Tomlinson, T.G. and Stride, G.O. (1945) Investigations into the fly populations of percolating filters. *Journal and Proceedings of the Institute of Sewage Purification*, pp. 140–148.

Tomlinson, T.G. and Williams, I.L. (1975) Fungi. In *Ecological Aspects of Used-Water Treatment, Vol. 1, The Organisms and their Ecology* (eds. C.R. Curds and H.A. Hawkes). Academic Press, London, pp. 931–952.

Tynen, M.J. (1966) A new species of *Lumbricillus* with a revised check list of the British Enchytraeidae (Oligochaeta). *Journal of the Marine Biological Association*, **46**, 89–95.

Unwin, D.M. (1981) A key to the families of British Diptera. *Field Studies*, **5**, 513–553.

Water Pollution Research Laboratory (1955) Report of the Director. HMSO, London.

Watson, W., Hutton, D.B. and Smith, W.S. (1955) Some aspects of gas liquor treatment on percolating filter beds. *Journal and Proceedings of the Institute of Sewage Purification*, pp. 73–85.

Weninger, G. (1964) *Jahreszyklus der Biozonose einser modernen Brockentropf Korperanlage.* Wasser und Abwasser, Beitrage zur Gerwasserf.

Weninger, G. (1971) Das auflereten kleiner Metazoen bei Abbauprozessen in ver- gleichender Siecht. *Sitz.-Ber. Ost. Akad. d. Wiss.* **179**, 129–158.

Wenjere, S. (2017) Assessment of the efficacy of BTI larvicide VectoBac® 12AS (BTI AM65-52) against "nuisance" flies water treatment plants in North West England. M.Sc. Thesis. School of Environment and Life Science, University of Salford, UK.

Wheatley, A.D. (1976) The ecology of percolating filters containing a plastic filter medium in relation to their efficiency in the treatment of domestic sewage. Ph.D. Thesis. University of Aston in Birmingham, UK.

Williams, I.L. (1977) Personal communication.

Williams, N.V. and Taylor, H.M. (1968) The effect of *Psychoda alternate* (Say.) (Diptera) and *Lumbricillus rivalis* (Levinsen) (Enchytraeidae) on the efficiency of sewage treatment in percolating filters. *Water Research*, **2**, (2), 139–150.

Williams, N.V., Solbe, J.F. de L.G. and Edwards, R.W. (1969) Aspects of the distribution, life history and metabolism of the Enchytraeid worms *Lubricillus rivalis* (Levinsen) and *Enchtraeus corornatus* (N. & V.) in a percolating filter. *Journal of Applied Ecology*, **6**, 171–183.

Wojnowska-Baryła, I., Cydzik-Kwiatkowska, A. and Zielinska, M. (2010). The application of molecular techniques to the study of wastewater treatment systems In *Bioremediation: Methods and Protocols. Methods in Molecular Biology* (ed. S. Cummings). Humana Press Inc., New York. US.

Woods, D.R., Williams, J.M. and Croydon, J. (1978) Fly nuisance control in treatment systems. *Water Pollution Control*, 77, 259–267.

Woombs, M. and Laybourn-Parry, J. (1986) The role of nematodes in low rate percolating filter sewage treatment works. *Water Research*, **20**, (6), 781–787.

Zakhvatkin, A.A. (1941) Fauna of U.S.S.R. Arachnoidea vol. I no. 1 Tyroglyphoidea (Acari). *Inst. Zool. Acad. Sci. Moscow*, **28**, 1–465.

Zhan, Y., Feng, Q., Haidong, Z., Zhang, Q. and Dong, B. (2013) Explosive propagation of *Aeolosomma hemprichi* in an activated sludge — Biofilm reactor. *Polish Journal of Environmental Studies*, **23**, (6), 2295–2302.

# 5
# Rotating Biological Contactors

## 5.1 The Process

Rotating biological contactors (RBCs) became commercially available in 1965, although the system was first developed in the late 1920s (Doman, 1929). They are now widely used throughout the world, and although particularly well suited for treating the wastewater from small communities, they are also used to treat large domestic and industrial loads. RBCs can be operated under aerobic, anoxic or anaerobic conditions and can be used for a wide variety of applications, including treating complex organic compounds and industrial wastewaters with constant innovations being developed (Chap. 9). However, in practice, they are almost exclusively used for aerobic secondary treatment, including nitrification, as well as being used for tertiary (i.e., polishing) treatment. The RBC is the intensification of the percolating filtration process, with the density of biomass on the discs reaching 200 g (DW) $m^2$, which is equivalent to a sludge loading of 40–60 kg MLVSS [mixed liquor volatile suspended solids] $m^{-3}$ in the activated process. When used for secondary treatment, RBCs require to be followed by a secondary settlement tank to remove biofilm and other solids from the final effluent, although sludge production is low compared to other secondary biological treatment processes.

RBCs are widely available as packaged plants under a variety of commercial names such as Biosurf™, Biodisc™, Biodrum™ and Biospiral™. The basic design consists of a series of flat or corrugated

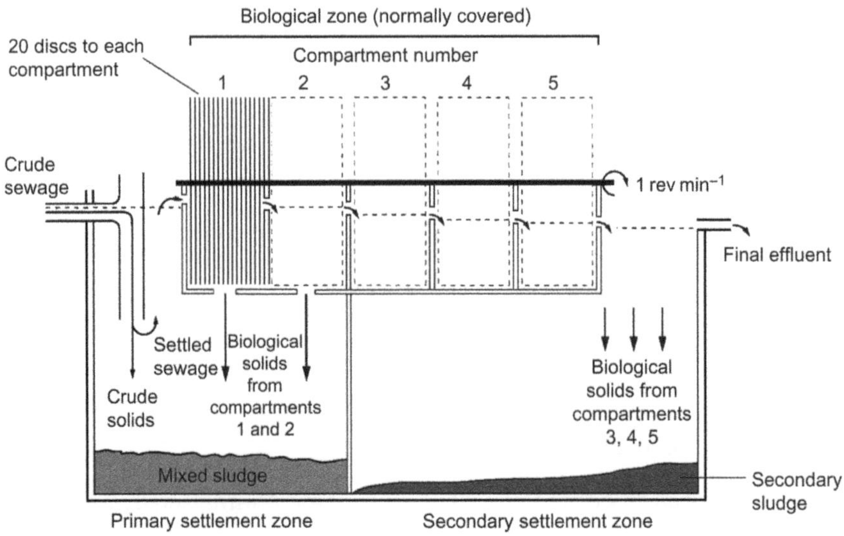


**Fig. 5.1.** Cross-section through a rotating biological contactor (Mann, 1979). Reproduced with permission of WRc plc.

discs that are 2.0–3.5 m in diameter, on which the biofilm develops, is mounted on a horizontal shaft and driven mechanically so that the discs rotate at right angles to the flow of settled wastewater (Fig. 5.1). The discs are usually plastic (MDPE), corrugated polythene, polyvinyl choloride (PVC), glass reinforced plastic (GRP), or expanded polystyrene, although other materials such as asbestos cement and expanded metal are also used. The surface area of the medium is normally between 150–200 $m^2m^{-3}$, with the higher surface area medium used at the outlet section for nitrification. Each disc is 10–20 mm thick and spaced about 20 mm apart (Figs. 5.2 and 5.3). The mounted discs are placed in a contoured tank that fits fairly closely to the rotating medium so that 35–40% of the disc area is immersed and are slowly but continuously rotated. As the discs rotate, wastewater enters the spaces between the discs when immersed and is then replaced with air when out of the liquid. In this way, a biofilm builds up on the discs in the same way as it develops on the medium in a percolating filter. Therefore, purification occurs with the biofilm alternately adsorbing organic nutrients from the wastewater and then obtaining oxygen from the atmosphere for oxidation. In practice, this mode of passive

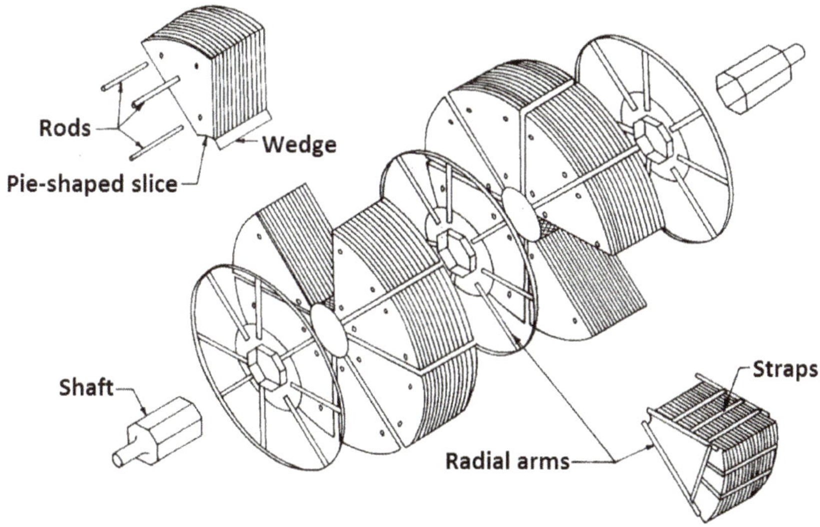


**Fig. 5.2.** Schematic view of a disc assembly in a RBC showing support rods — used in many designs — that ensure exact spacing between discs (IWEM, 1998). Reproduced with permission of the Chartered Institution of Water and Environmental Management.

**Fig. 5.3.** Old rectangular activated sludge tanks can be upgraded to RBC units. An RBC unit is being lowered into an activated sludge tank as a retrofit operation. The support rods are necessary when larger discs are used. Reproduced with permission of System S&P GmbH (https://www.system-s-and-p.de/).

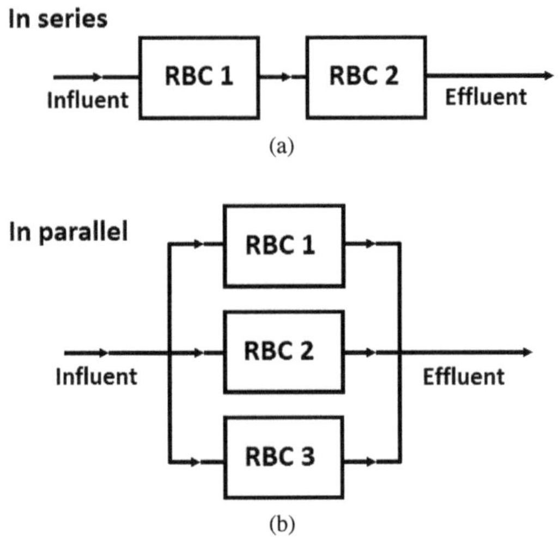


**Fig. 5.4.** Rotating biological contactors can be operated in series or in parallel to give flexible operation.

oxygen transfer is not very efficient and problems of anaerobic conditions developing within the first stage of the unit are possible at high organic loadings. This can be overcome by recirculation, using units in parallel (Fig. 5.4) or better loading control by employing an equalization tank before the RBC. At immersion depths >40%, supplementary air may be required to maintain aerobic conditions. The redox potential throughout the biofilm can be controlled by the depth of immersion of the discs that controls oxygen availability, creating anoxic and/or anaerobic zones with a depth that allow other oxidative reactions to occur, including denitrification and sulfate reduction (Alleman *et al.*, 1982) (Chap. 7).

Discs are usually but not exclusively used as support media. Other configurations, such as lattice constructions or wire mesh containers filled with random plastic media, have also been used successfully, with the shaft, either driven by air or water passing through the media. The flow of wastewater through the tank and the action of the rotating medium produce a high hydraulic shear on the biofilm that ensures efficient mass transfer from the liquid into the biofilm, and at the same time, controls excessive biofilm accumulation. Discs are arranged in groups separated by baffles to minimize short-circuiting and the effect of surges of flow. Normally, a minimum of four compartments, separated by baffles, are used to simulate plug flow with a small head loss of between 10–20 mm between each compartment. With large installations, plug-flow conditions are achieved by having several complete RBC units in series (Steels, 1974) (Figs. 5.4 and 5.5). The system consumes very little energy as the discs are evenly balanced along the drive shaft, thereby requiring only a small motor to drive the rotor, which results in low-power consumption for the amount of biochemical oxygen demand (BOD) removed. For example, in a 300 PE unit, the power consumption of the motor is only 0.3 kW. RBC units are invariably covered, with small units covered with a molded plastic or GRP cover (Fig. 5.6) and larger units housed in buildings. Covering the media is important to protect the exposed biofilm on the discs from the weather, especially frost and wind, which can damage the film. Wind can also increase the load on the motor driving the rotor. Covering also insulates the system and reduces heat loss, thereby increasing the rate of oxidation. By insulating the cover, such units can operate successfully, even in arctic

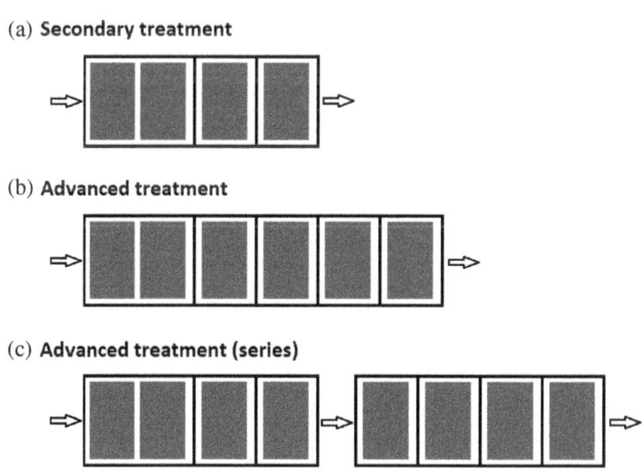


**Fig. 5.5.** Different configurations of the compartments within an RBC. The more compartments there are, the more specialized the biofilm community becomes. Compartments can also be operated as anoxic zones for pre and post denitrification (Chap. 7).

**Fig. 5.6.** Small RBC showing separate disc compartments. The hood is made of glass-reinforced plastic and can be raised for inspection.

conditions. Excluding sunlight not only prevents UV damage to the biofilm and plastic medium, but it also prevents algal development on the biofilm. The algae can reduce overall performance and create a dense strong biofilm that is less likely to slough off at the normal rotation speed but also puts extra loading in terms of weight on the shaft and motor. The cover also eliminates fly nuisance and even controls odors to some extent, although, where fine-mesh grills are used they must be regularly cleaned to ensure adequate air movement in and out of the unit. Using green covers makes such plants inconspicuous, especially as they have the same head loss as a septic tank (<100 mm), and therefore are low-form structures, making them ideal for use in open areas where landscape quality is important, for example, golf clubs, restaurants, hotels and general amenity areas.

Separation of the discs varies between 15–30 mm, which is a compromise between reducing the spacing to maximize surface area per meter length of the rotor, while preventing the biofilm from bridging the gap between discs, so reducing the effective surface area and

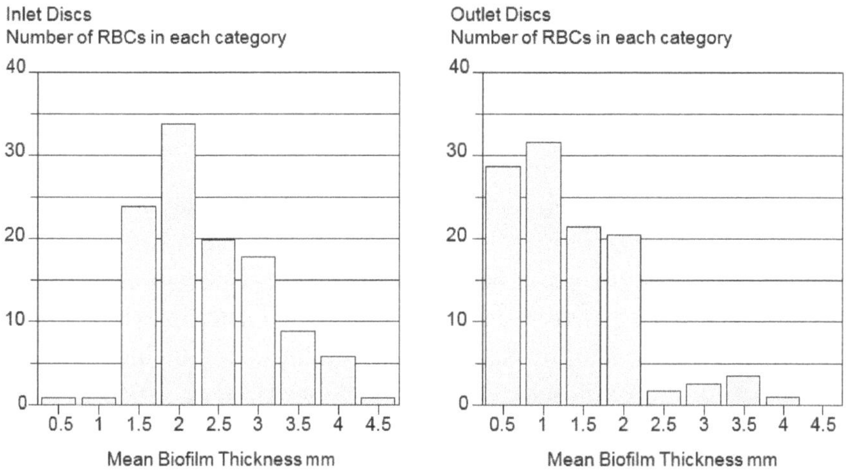


**Fig. 5.7.** Frequency distribution of mean biofilm thickness on the inlet and outlet discs at 114 rotating biological contactor sites (Griffin and Finday, 2000). Reproduced with permission of IWA Publishing.

leading to low dissolved oxygen within the film. This is especially a problem where shearing velocities are too low to dislodge excess film. A survey conducted on 114 RBCs by Severn Trent Water showed biofilm thickness to vary between 0.5 and 4.5 mm, with significantly thicker films developing on inlet discs compared to those at the outlet (Fig. 5.7) (Griffin and Findlay, 2000). Allowing a minimum spacing of 15 mm for water to pass between discs and 5 mm and 3 mm for maximum permissible biofilm development on the inlet and outlet discs, respectively, they used 25 mm and 21 mm for the inlet and outlet disc separation distances, respectively. The thickness of the biofilm is managed by preventing organic overloading. Using these spacings, the approximate surface area for a 1.0 m diameter disc unit is 50–80 $m^2$ per meter of rotor length, rising to 700 $m^2m^{-1}$ for a 3.0 m diameter disc unit. The average depth of the disc immersion was 40%.

## 5.2 Ecology of RBCs

The biofilm that develops on the discs is a complex and diverse community of bacteria, filamentous bacteria, protozoa and metazoa.

There is a gradation of microbial species along the reactor path, with the first discs developing a heavy growth of heterotrophic bacteria and fungi. The structure of the biofilm is similar to that in percolating filters, with the initial development following the main phases of attachment and development controlled by quorum sensing (Sec. 2.2). As the biofilm develops, the pores enlarge, allowing deeper penetration of nutrients and the exchange of gases, with young biofilms having numerous fine pores in contrast to fewer wider pores in older biofilms, increasing mass transfer (Zahid and Ganczarczyk, 1994). The density of the biofilm increases with depth, due to increases in cell density and retention of mineral and recalcitrant organic material on available adsorption sites of the EPS and bacterial walls, causing the rate of diffusion to decrease as the surface of the medium is approached, leading to a loss of mass transfer potential and then to cell starvation and anoxia. Filamentous bacteria, especially *Sphaerotilus natans*, *Nocardia* spp. and *Beggiatoa* are commonly recorded. While both are indicative of heavy organic loading, *Beggiatoa* oxidizes reduced sulfur compounds, so while its presence at the inlet is normal, plant overloading occurs if it extends through several stages or becomes excessively luxuriant. The bacterium forms a smooth but dense white layer over the existing biological biofilm and can become quite thick. It contrasts dramatically to the normal thin chocolate-brown biofilm that is normally found on the majority of the discs and so the extensive white growth often causes concern to operators when discs are periodically inspected. *Beggiatoa* outcompetes the normal heterotrophic non-filamentous bacteria for space and oxygen, resulting in very heavy biofilm growths, which adds to the mechanical stress on the RBC shaft and bearings, as well as significantly reducing the BOD removal efficiency. Below the *Beggiatoa* film is a black layer (the discoloration being caused by ferrous sulfide precipitation), which is rich in *Desulfovibrio* spp. and are sulfate-reducing bacteria. There is an interesting relationship between *Beggiatoa* and *Desulfovibrio* (Fig. 5.8). In the inner anaerobic layer, fermentative bacteria provide the end products used by sulfate-reducing bacteria (e.g., organic acids, alcohols). *Beggiatoa* in the outer aerobic zone uses the hydrogen sulfides, produced by the sulfate-reducing bacteria in the lower anaerobic zone, as an electron donor oxidizing it to elemental sulfur

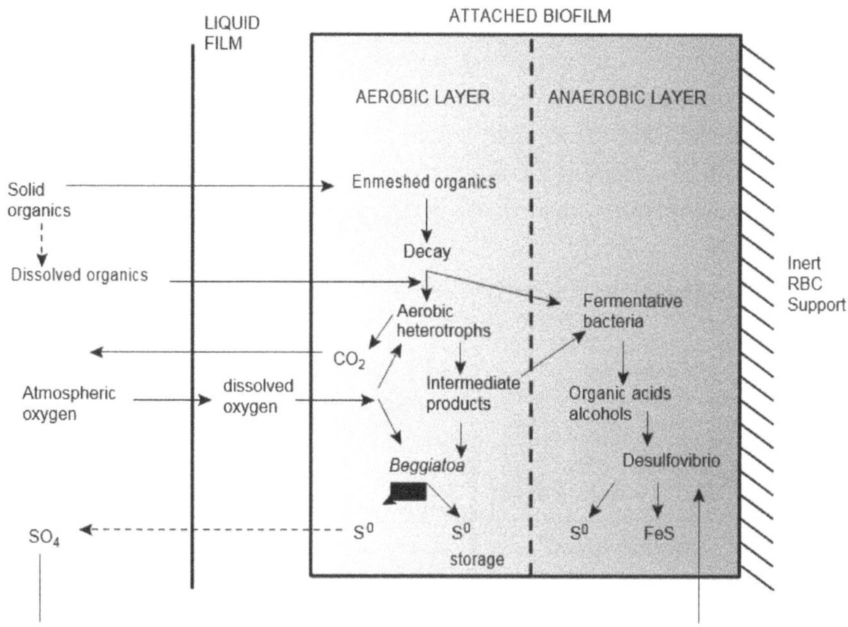


**Fig. 5.8.** Relationship between the bacteria *Beggiatoa* and *Desulfovibrio* in the biofilm of RBCs (Bitton, 1999). Reproduced with permission of John Wiley and Sons Inc.

(Alleman *et al.*, 1982). A succession of protozoan species is seen on successive discs following the saprobic rating of species (Liebmann, 1951). These can be used as indicators of the state of organic loading and RBC effluent quality (Kinner *et al.*, 1988). This succession of species is similar to that seen in the plug-flow activated sludge processes (Kinner and Curds, 1989):

> Bacteria colonization → flagellates → amoebae → free swimming bacteriovorous ciliates (e.g., *Colpidium* spp.) → nematodes → stalked ciliates (e.g., *Vorticella* spp.) → rotifers.

## 5.3 Operational Control

The speed of rotation is limited by the shearing of the biofilm at the periphery of the discs, but the normal range is between 0.5–10 rpm, typically 0.5–1.5 rpm depending on the diameter of the disc. Biofilm is sloughed off continuously and will remain in the liquid phase, thus adding to the overall treatment process. Eventually, sloughed biofilm is carried away to a separate sedimentation tank, although in smaller units the secondary settlement tank is built into the chamber. Most units also have a digestion chamber below the discs to reduce the load to the secondary settlement tank, thereby reducing the overall quantity of sludge produced and the frequency of desludging required. The collected solids in the digestion chamber undergo anaerobic digestion (Fig. 5.9), with desludging normally required every 1–3

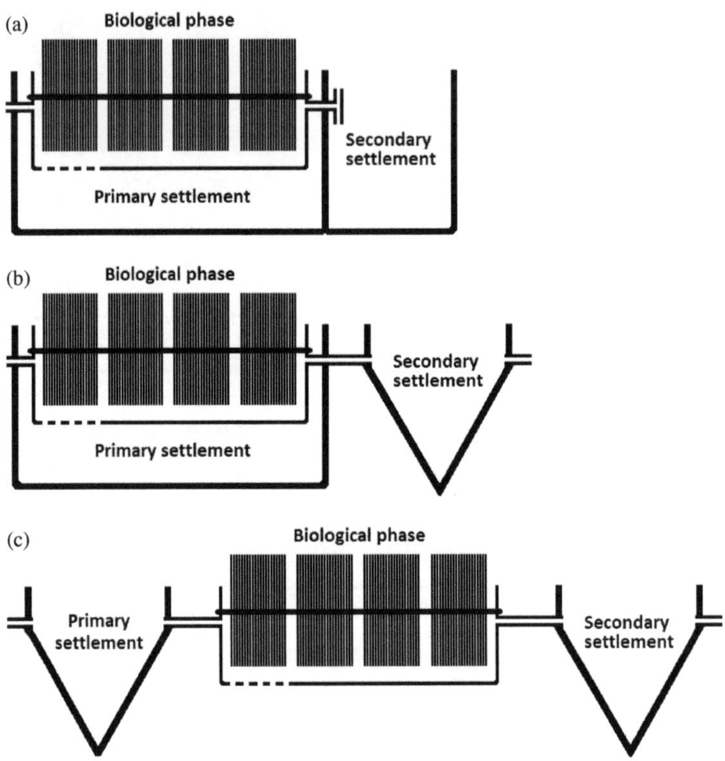


**Fig. 5.9.** Some RBC configurations: (a) Primary and secondary settlement are incorporated into a single unit. A common design for packaged systems used for individual houses and small commercial premises, (b) a primary settlement incorporated into the main RBC unit with a separate secondary settlement tank. This design is used at commercial premises, and (c) standard wastewater treatment plant layout for larger populations where primary and secondary settlement are separate unit processes.

months. Failure to maintain a regular desludging program results in odor production. Modern designs do not incorporate digestion chambers, which are now only seen in small units for single dwellings.

An electric motor is not always necessary to drive the rotor and it is possible to use air-induced drive systems. For example, air is sprayed into the rotating medium from below and to one side of the horizontal shaft. The liquid movement and asymmetric buoyancy from air collecting in one side of the medium induce it to rotate (Winkler, 1981). This method reduces the risk of anaerobiosis in the first compartment or in the first RBC unit of a multiple unit system used for large populations. It is also used to uprate the activated sludge process. The disc system can be placed in an existing activated sludge aeration tank to significantly increase the biomass concentration within the tank, without the need to increase the aeration capacity to satisfy the oxygen demand of an increased biomass (MLSS) concentration (Guarino *et al.*, 1980). This led to the development of the integrated fixed-film activated sludge (IFAS) systems of today (Sec. 6.6.2).

Small plants are designed on the basis of BOD loading per unit area of discs using both sides of each disc in the calculation. Bruce and Merkens (1975) found that a Royal Commission effluent of 20 mg BOD $L^{-1}$ and 30 mg suspended solids $L^{-1}$ (20:30) could be produced at organic loading rates of <6 g BOD $m^{-2}d^{-1}$. A conventional low-rate percolating filter containing 50 mm of mineral medium and loaded at 0.10–0.12 kg BOD $m^{-3}d^{-1}$ can be loaded at a surface loading of 1.0–1.2 g BOD $m^{-2}d^{-1}$, which is five or six times less than an RBC, to produce a similar quality effluent (Pike, 1978). Higher loading figures have been given for 80–90% BOD reductions, ranging from 6–20 g BOD $m^{-2}d^{-1}$, although the organic loading can be increased by a factor of 10 for efficient partial treatment (Hao and Hendricks, 1975a; 1975b). Loadings to larger units are based on hydraulic loading per unit area of the disc. The calculations of loadings are complex and explained in detail by Steels (1974) and Wilson (1981). For obvious reasons, manufacturers feel that loadings to RBCs are too conservative, thus making them less attractive when compared to the activated sludge process. The calculations of loadings have been reviewed by Lumbers (1983). The recommended loading for RBCs to give a 20:30 effluent should not exceed 5 g BOD $m^{-2}d^{-1}$ (BSI, 1983). In practice, designs loading are significantly lower than this at 4 g BOD $m^{-2}d^{-1}$ in order to achieve a consistent 20:30 effluent. Experience has shown that if full nitrification is required, then loading should not exceed 2.5 g BOD $m^{-2}d^{-1}$. These values refer to the total medium surface area of the RBC unit; however, if the organic loading to the first stage exceeds 31 g BOD $m^{-2}d^{-1}$, then problems with excessive *Beggiatoa* development will follow (Brenner *et al.*, 1984). A minimum hydraulic retention time (HRT) of 0.6–1.5 hours is required for carbonaceous oxidation and 1.5–3.8 hours for nitrification (Wang *et al.*, 1984), although Cheung and Mack (1982) showed that an HRT of over 3 hours did not improve carbonaceous removal. Maximum flows to RBC units are restricted to 6 DWF by the provision of a storm sewage overflow, often linked to a dedicated stormwater reed bed. When pumped from a holding tank, wastewater should be applied to the RBC unit little at a time but often. This is done by using a simple timer control that starts the pump. For example, for only 1 to 2 minutes every hour, however, timed loading must be overridden by surface-level controls to prevent flooding of the unit. A minimum dissolved oxygen concentration of 1.5–2.0 mg $L^{-1}$ is required to ensure maximum BOD removal and to minimize nuisance. The dissolved oxygen concentration in the wastewater is a function of the oxygen utilization by the biofilm and the oxygen transfer between the biofilm and the liquid. Where the organic loading is too high, the oxygen uptake rate of the biofilm exceeds the oxygen transfer rate, resulting in low dissolved oxygen concentrations, loss of performance and development of *Beggiatoa*.

Depending on the organic loading, nitrification can occur on the first group of discs, although nitrifying activity is usually limited. Like percolating filters, nitrifying bacteria in RBCs are found toward the end of the system on the discs in the third and fourth compartments of a four-compartment unit (Okabe *et al.*, 1996). Due to the plug flow nature of the system, there is a tendency for different nitrifier groups to occur depending on substrate availability, with ammonium oxidizing bacteria (AOB) occurring on the discs releasing nitrite that support the development of nitrite oxidizing bacteria (NOB) further along the plug flow system. The biofilm in nitrifying RBCs, which are carbon limited, is comprised of 50% ammonia and nitrite-oxidizing bacteria (AOB and NOB), with the remainder heterotrophs using soluble microbial products from the biofilm endogenous breakdown. There is evidence that specific heterotrophic groups utilize different breakdown products, creating dependent community relationships (Okabe *et al.*, 2005). There is a net loss of nitrogen through the system that approaches 20% because of denitrification at the media-film interface (Ellis and Banaga 1976). Some designs incorporate recirculation of the final effluent back to the first stage. The idea behind this is to supply combined oxygen in the heavily loaded first stage to prevent septicity and to encourage denitrification to take place. The nitrogen oxidation rate is similar to that found in filters at approximately 1 g $NH_3$-N $m^{-2}d^{-1}$ at 20°C (Antonie, 1976). Major variables affecting nitrification include total disc area, wastewater temperature, influent nitrogen, and BOD concentrations and flow rate (Barnes and Bliss, 1983). For example, based on the Van't Hoff–Arrhenius relationship, nitrification at 5°C may require 2.5 times more medium surface area than at 13°C. Examples of RBCs used for denitrification are given by Cortez *et al.* (2011) and Courtens *et al.* (2014). Unlike simultaneous nitrification and denitrification, biological phosphorus removal is difficult to achieve in RBCs due to the problems of creating the ideal oxic-anaerobic step required for the adequate and sustained development of phosphorus accumulating organisms (PAO) (Hassard *et al.*, 2015; Yun *et al.*, 2004). This is considered further in Chap. 7.

RBCs have many advantages over the percolating filter process. Some examples are: complete wetting of the media, no clogging of the media due to excessive biofilm accumulation reducing the retention time (in practice the retention time of an RBC is the same as a percolating filter), and regular exposure to the air ensures unlimited oxygen and the area of land used is only 10% of that required by the equivalent treatment capacity supplied by low-rate filtration. RBCs are available as package units and can be installed very quickly with integral primary and secondary settlement compartments. When designed correctly, inspection periods can be extended to several weeks, thus making them ideal for remote locations that may be cut off for short periods during winter. Other advantages include low sludge production, excellent process control, ease of operation, a high degree of BOD removal including substantial nitrification (Barnes and Bliss, 1983), good settleability of solids, low power consumption, no distribution problems, no fly problems, and no recirculation is required to ensure wetting of the biofilm, as is often necessary with small percolating filters. The low-power requirements have also attracted interest in running them on solar-charged batteries, making them close to carbon neutral.

There are some potential disadvantages, which include frequent motor drive and bearing maintenance. Also, the process is affected by surges of flow and is sensitive to overloading. Power failure causes a total loss of efficiency, and where the discs are left standing in the wastewater for any length of time, uneven biofilm distribution results with thicker biofilm developing on the submerged surfaces and the biofilm on the exposed surfaces drying out. This causes a weight imbalance across the discs and this causes uneven rotation of the shaft (i.e., loping), leading to severe motor wear, which results in eventual failure. While the active depth of biofilm on the discs is <600 μm and 200 μm for nitrifying film, thick growths give rise to anaerobic conditions at the film-medium boundary. Such heavy biofilm development results in odor when the biofilm sloughs and also in mechanical problems due to the excessive weight, such as collapsed medium, failed bearings and drive mechanisms, and also broken shafts. Biofilm thickness is directly related to organic loading but is also a function of the shear forces acting on the biofilm (Möhle and Langemann, 2007). Excessive biofilm development is corrected by increasing the speed of rotation, which causes the shear forces on the biofilm to increase and eventually causing it to be hydraulically scoured off. Finally, RBCs are comparatively expensive compared to other biological units, especially for small treatment plants. But their inconspicuousness, lack of odor, low fly nuisance and quietness make them ideal where treatment must be done close to houses.

RBCs have been used for secondary treatment for populations between 50 and 2,000 PE by Severn Trent Water since 1989. Since then over 335 units have been constructed on 223 sites. Griffin and Findlay (2000) outlined their design and operation. The RBCs use high-density polyethylene (HDPE) discs that are submerged 40% of their diameter and rotated between 0.5–2.0 rpm, although 1 rpm is standard. The RBC design is particularly robust with the disc unit consisting of two mild steel end frames connected by through rods that support the discs. At least three separate compartments are required in the biozone to promote plug flow characteristics. To avoid excessive biofilm thickness and nuisance organisms, such as *Beggiatoa* in the first compartment, a maximum loading of 20 g BOD $m^{-2}d^{-1}$ is specified. The maximum specific surface area of the medium is 150 $m^2m^{-3}$ in the first compartment and 220 $m^2m^{-3}$ in the remaining compartments. Performance is linked directly to disc loading (Fig. 5.10), with 4 g BOD $m^{-2}d^{-1}$ used where a consent of >15 mg $L^{-1}$ ammoniacal nitrogen is set, reducing to 2.5 BOD $m^{-2}d^{-1}$ where it is <10 mg $L^{-1}$. This lower loading has been found to meet the 95th percentile ammonical N standard of 3 mg $L^{-1}$ that is currently set for sensitive receiving waters. Long term monitoring has shown that the RBCs

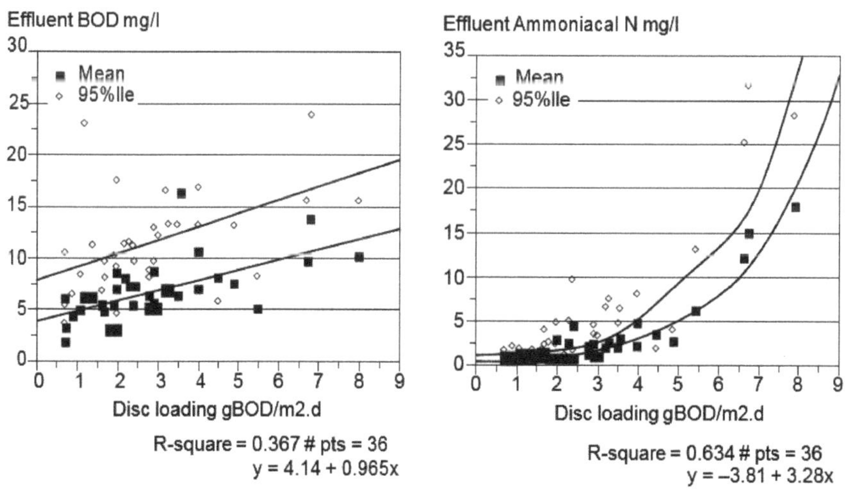


**Fig. 5.10.** Annual mean and 95th percentile concentrations for BOD and ammoniacal nitrogen at different disc loadings (g $BOD_5$ $m^{-2}d^{-1}$) for 26 rotating biological contactor sites (Griffin and Finday, 2000). Reproduced with permission of IWA Publishing.

used by Severn Trent Water do not reliably achieve a 95th percentile compliance with BOD standards of <25 mg $L^{-1}$. Therefore, they have adopted a strategy of using tertiary treatment reed beds that can produce final effluents with a BOD of <3 mg $L^{-1}$, suspended solids of <4 mg $L^{-1}$, and fully nitrified (Upton *et al.*, 1995). RBCs have been reviewed in detail by Cortez *et al.* (2008), Patwardhan (2003) and Hassard *et al.* (2015).

## References

Alleman, J., Veil, J. and Canaday, J. (1982) Scanning electron microscope evaluation of rotating biological contactor film. *Water Research*, **16**, 543–550.

Antonie, R.L. (1976) *Fixed Biological Surfaces — Wastewater Treatment. The Rotating Biological Contactor.* CRC Press Inc, Cleveland, Ohio.

Barnes, D. and Bliss, P.J. (1983) *Biological Control of Nitrogen in Wastewater Treatment.* Spon, London.

Bitton, G. (1999) *Wastewater Microbiology, 2nd ed.* John Wiley and Sons, New York, US.

Brenner, R.C., Heidman, J.A., Opatken, E.J. and Petrasek, A.C. (1984) Design information on rotating biological contactors. EPA-600/2-84-106. US Environmental Protection Agency, Cincinnati, Ohio.

Bruce, A.M. and Merkens, J.C. (1975) Developments in sewage treatment for small communities. *Proceedings of the 8th Public Health Engineers Conference*, Loughborough University, pp. 65–96.

BSI (1983) Design and installation of small sewage treatment works and cesspools. Code of Practice: BS 6297. British Standards Institution, London.

Cheung, P.S. and Mack, G. (1982) The influence of organic loading and retention period on the performance of the rotating disc system. *Journal of the Water Pollution Control Federation*, **18**, (4), 553.

Cortez, S., Teixeira, P. and Mota, M. (2008) Rotating biological contactors: A review on main factors of performance. *Reviews in Environmental Science and Bio/Technology*, 7, 155–172.

Cortez, S., Teixeira, P., Oliveira, R. and Mota, M. (2011) Denitrification of landfill leachate with high nitrate concentration in an anoxic rotating biological contactor. *Biodegradation*, **22**, 661–671.

Courtens, E.N.P., Boon, N., De Clippeleir, H., Berckmoes, K., Mosquera, M., Seuntjens, D. and Vlaeminck, S.E. (2014) Control of nitratation in an oxygen-limited autotrophic nitrification/denitrification rotating biological contactor through disc immersion level variation. *Bioresources Technology*, **155**, 182–188.

Doman, J. (1929) Results of operation of experimental contact filter process with partially submerged rotating plates. *Sewage Works Journal*, **1**, 555–560.

Ellis, K.V. and Banaga, S.E. (1976) A study of rotating disc treatment units operating at different temperatures. *Water Pollution Control*, **75**, 73–91.

Griffin, P. and Findlay, G.E. (2000) Process and engineering improvements to rotating biological contactor design. *Water Science and Technology*, **41**, (1), 137–144.

Guarino, C.F., Nelson, M.D., Lazanof, M. and Wilson, T.E. (1980) Uprating activated-sludge plants using rotary biological contactors. *Water Pollution Control*, **79**, 255–271.

Hao, O. and Hendricks, G.F. (1975a) Rotating biological reactors remove nutrients: Part I. *Water and Sewage Works*, **122**, (10), 70–73.

Hao, O. and Hendricks, G.F. (1975b) Rotating biological reactors remove nutrients: Part II. *Water and Sewage Works*, **122**, (11), 48–50.

Hassard, F., Biddle, J., Cartmell, E., Jefferson, B., Tyrrel, S. and Stephenson, T. (2015) Rotating biological contactors for wastewater treatment — a review. *Process Safety and Environmental Protection*, **94**, 285–306. https://doi.org/10.1016/j.psep.2014.07.003

Kinner, N.E. and Curds, C.R. (1989) Development of protozoa and metazoan communities in rotating biological contactor biofilms. *Water Research*, **23**, 481–490.

Kinner, N.E., Curds, C.R. and Meeker, L.D. (1988) Protozoa as indicators of effluent quality in rotating biological contactors. *Water Science and Technology*, **20**, (11–12), 199–203. https://doi.org/10.2166/wst.1988.0285

Liebman, H. (1951) *Handbuch der Frishwasser-und Abwasser-biologie.* Gustav Fischer, Jena.

Lumbers, J.P. (1983) Rotating biological contactors: Current problems and potential developments in design and control. *Public Health Engineering*, **11**, 41–45.

Mann, H.T. (1979) Septic tanks and small treatment plants. Technical Report 107. Water Research Centre, Medmenham, UK.

Möhle, R. and Langemann, T. (2007) Structure and shear strength of microbial biofilms as determined with confocal laser scanning microscopy and fluid dynamic gauging using a novel rotating disc biofilm. *Biotechnology and Bioengineering*, **98**, 747–755.

Okabe, S., Hiratia, K., Ozawa, Y. and Watanabe, Y. (1996) Spatial microbial distribution of nitrifiers and heterotrophs in mixed population biofilms. *Biotechnology and Bioengineering*, **50**, 24–35.

Okabe, S., Kindaichi, T. and Ito, T. (2005) Fate of 14C-labeled microbial products derived from nitrifying bacteria in autotrophic nitrifying biofilms. *Applied Environmental Microbiology*, **71**, 3987–3994.

Pateardhan, A.W. (2003) Rotating biological contactors: A review. *Industrial Engineering and Chemical Research*, **42**, 2035–2051.

Pike, E.B. (1978) The design of percolating filters and rotating biological contactors, including details of international practice. Technical Report TR93. Water Research Centre, Stevenage.

Steels, I.H. (1974) Design basis for the rotating disc process. *Effluent and Water Treatment Journal*, **14**, 431–445.

Upton, J.E., Green, M.B. and Findlay, G.E. (1995) Sewage treatment for small communities: The Severn Trent approach. *Journal of the Chartered Institution of Water and Environmental Management*, **9**, (1), 64.

Wang, K., Wang, M.H. and Poon, C.P.C. (1984) Rotating biological contactors. *Effluent and Water Treatment Journal*, **24**, (3), 93.

Wilson, F. (1981) *Design Calculations in Wastewater Treatment.* Spon, London.

Winkler, M.A. (1981) *Biological Treatment of Wastewater*. Ellis-Horwood, Chichester.

Yun, Z., Lee, H. and Choi, E. (2004) Enhanced biological phosphorus removal in RBC with SBR modification. *Water Science and Technology*, **50**, (10), 121–130.

Zahid, W and Ganczarczyk, J. (1994) Fractal properties of the RBC biofilm structure. *Water Science and Technology*, **29**, (10), 271–279.

# 6
# Submerged Fixed-Film Reactors

## 6.1 Introduction

One of the limiting factors to increase the efficiency of percolating filtration is the limited surface area available for biofilm development without reducing the size of the interstices so that they become blocked with biofilm growth and thus restrict flow and aeration. While the introduction of plastic media improved the situation, the total area of fixed active biomass remains much lower in percolating filters than in the activated sludge process per unit volume of the reactor. In the 1980s, the use of very fine media, such as expanded clay nodules or even sand, and the passing of wastewater upward through the media at a sufficient velocity to fluidize the media was proposed (Sec. 6.2). In this way, the interstices could not become blocked no matter how much biomass developed on the medium. The small media offered specific surface areas of up to 3,500 $m^2m^{-3}$, allowing biomass concentrations equivalent to a mixed liquor suspended solid (MLSS) concentration of 30–40,000 mg $L^{-1}$. The use of such fine media providing very high specific surface areas allowed more compact and versatile systems to be designed. From this original concept, two further submerged systems emerged — biological aerated flooded filters (BAFFs), otherwise known as biological aerated filters (BAFs) (Sec. 6.3), and submerged aerated filters (SAFs) (Sec. 6.4). Both systems use a submerged solid medium with oxygen provided by diffusers or other mechanical devices. There is considerable confusion between the two systems, and the terms are often used interchangeably. This is because there is such a wide variation between individual systems that either term is often not a sufficiently accurate descriptor. However, most authors, and the definition used in this text, define BAFF technology as filters that achieve solids separation by backwashing, while SAF technology is filters that require a settlement tank for solids separation (Ahmadi *et al.*, 2014).

With the intensification of wastewater treatment and the need to make processes more efficient conventional systems have come under increasing criticism. For example, percolating filters are generally limited in terms of hydraulic and organic loadings and are prone to ponding due to excessive biofilm accumulation, which leads to channeling and loss of performance as well as potential odor and fly problems (Chap. 3). Rotating biological contactors (RBCs) can be susceptible to mechanical failure and are also expensive when used for larger installations (Chap. 5). While many of these problems were overcome through the development of submerged systems, these also have limitations. For example, fluidized bed reactors (FBRs) are difficult to operate (Sec. 6.2), granular media reactors require backwashing (Sec. 6.3), and the general problem in fixed media systems in ensuring all the media surface is used with excessive biofilm development, which causes blockages and channeling (Sec. 6.4) (van Haandel and van der Lubbe, 2012). These limitations have driven research into a new range of submerged reactors based around the concept of the moving bed biofilm reactor (MBBR) and the integrated fixed-film activated sludge (IFAS) process.

## 6.2 Fluidized Bed Reactors

To provide sufficient surface area to act as support for biofilm development in a fixed-film reactor (FFR) system, a medium is used. However, when 50 mm blast-furnace slag is used, 48.7% of the volume of the filter is taken up with the medium itself. This falls to just 8.7% for the random plastic medium Flocor RC™. In FBRs, the problem of providing sufficient fixed biomass for complete treatment is overcome by using a much smaller medium than normal, such as sand or finely crushed glass. By passing the wastewater through such a medium in an upward direction at sufficient velocity, the medium becomes fluidized. Sand provides a large increase in specific surface area compared to other media (~3,300 $m^2m^{-3}$), which allows considerable biomass to develop, which is equivalent to an MLSS concentration of up to 40,000 mg $L^{-1}$. In addition, the fluidization of the sand particles prevents clogging or excessive biofilm accumulation, as the interstices are neither fixed nor stable. This mode of operation allows the density of the biomass to be maintained at high concentrations without having to recycle solids as in the activated sludge process. Also, the quantity of biomass is controllable, so there are no wide seasonal fluctuations in biomass concentration, as seen in other FFR types.

The basic layout of FBRs is shown in Fig. 6.1. Classification of fluidized beds is confusing as they are used for a wide range of industrial phsico-chemical processes, and it was not until the 1970s that they were used for the biological treatment of wastewaters. The expansion of the medium (solid) is normally achieved using the wastewater (liquid), known as a liquid-solid or two-phase system. Alternatively, expansion of the medium can be achieved using both wastewater and air (or oxygen) together (gas), a gas-liquid-solid or three-phase system. Two-phase systems are much simpler to operate and are the most common form of fluidized reactors used in wastewater treatment, although three-phase systems offer more flexibility in operation, especially when treating more complex wastes.

The high biomass concentration in FBRs results in a very high oxygen demand that can only be satisfied by injecting pure oxygen into the influent wastewater stream as it enters the base of the reactor, giving an oxygen concentration approaching 100 mg $L^{-1}$. The even distribution of the influent wastewater into the medium is achieved by injecting the wastewater downward into a conical base (Cooper and Wheeldon, 1982). The expansion of the bed within the reactor is controlled by the rate of wastewater input, so the effluent can be removed from the reactor below the level of the expanded medium without a loss of media. A secondary sedimentation tank can be supplied but is not necessary as the biomass can be retained within this up-flow system by careful operation. The sand-covered biomass is regularly removed,

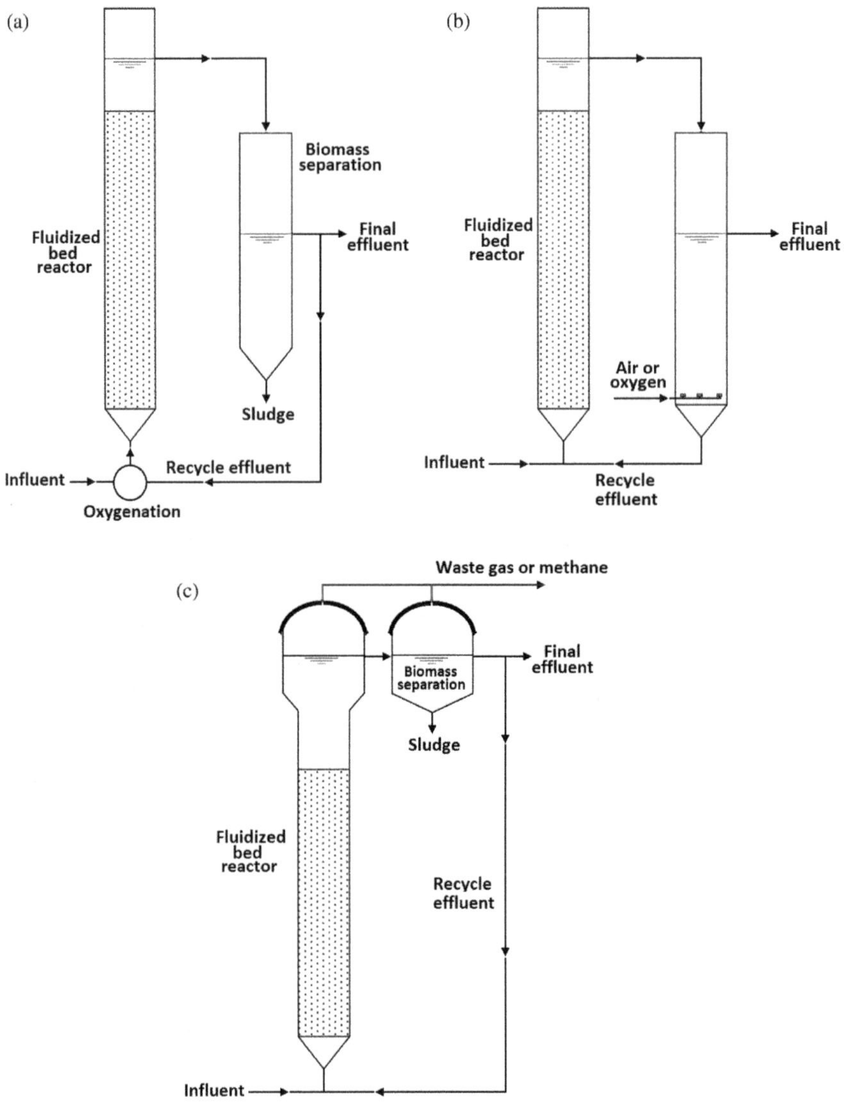


**Fig. 6.1.** Common designs of fluidised bed reactors used for wastewater treatment. (a) and (b) aerobic two-phase reactor systems, and (c) anaerobic fluidized bed two-phase reactor (AFBR) system.

and the biomass separated from individual sand particles by passing it through a hydroclone where a high shear strips away any attached microbial growth. The clean sand is subsequently returned to the reactor with the separated biomass further treated along with the primary sludge before disposal. Conventional FBRs are ideal for treating high strength wastewaters and are particularly useful for treating industrial wastewaters where the loading is variable (Cooper and Wheeldon, 1980; 1982). They operate at short retention times, with 95% removal of the biochemical oxygen demand (BOD) achieved within 16 minutes. When operated as a nitrifying process, a 20 mg $L^{-1}$ concentration of ammonia-nitrogen can be reduced by 99% in 11 minutes at 24°C, which represents an ammonia-nitrogen loading of 900 mg $m^{-2}d^{-1}$ at a hydraulic loading of 140 $m^3m^{-3}d^{-1}$. This is similar to that of a standard nitrifying filter (Chap. 7) (Jeris *et al.*, 1977).

Fluidized beds are compact systems compared to other FFRs or the activated sludge process, resulting in a lower land requirement and capital cost. However, operating costs are significantly greater because high purity oxygen and pumping are both required. The sludge is more concentrated at 10% of dry solids compared with 2% of gravity-thickened activated sludge. There are no odors or fly nuisances either, unlike conventional FFR designs. The fluidized bed can be operated aerobically, anoxically or anaerobically. Anaerobic beds tend to expand rather than fluidize and are used for the conversion of carbonaceous wastes into gaseous end products. Therefore, these reactors require a gas/liquid separation stage (Fig. 6.1c). Anaerobic beds should be operated in the mesophilic temperature range so heating is required (Moteleb *et al.*, 2001). Anoxic beds have been also developed specifically for denitrification (Gauntlett and Craft, 1979).

Although sand (0.2–2.0 mm in diameter) is the most common support media in aerobic systems, anthracite, carbon and glass are also used. Small particles have a higher specific surface area but a lower settling velocity than large particles. Therefore, larger fabricated media have been developed, which are porous and allow the biomass to grow within the media as well as on the outer surface such as reticulated polyester foam pads (25 × 25 × 10 mm). There are two full-scale aerobic fluidized beds on the market. Oxitron™ is manufactured by

Dorr-Oliver Inc. and uses sand particles 250–550 mm in diameter with a specific surface area of >3,000 $m^2m^{-3}$. It can support a large biomass of between 10–17 kg MLVSS [mixed liquor volatile suspended solids] $m^{-3}$, with the excess biomass removed by shear. Oxygen is supplied by dissolving high purity oxygen into the influent wastewater stream (Fig. 6.2) (Sutton *et al.*, 1981; Hoyland and Robinson, 1983). The CAPTOR™ process is the original commercially available system that employed plastic foam pads. Manufactured by Simon-Hartley Ltd, the excess biomass is removed by passing the pads through a compression roller, which releases a concentrated sludge with the pads returned to the reactor with a reduced microbial population. Oxygen is supplied by using a proprietary oxygenator (Walker and Austin, 1981). CAPTOR™ is also operated as an IFAS system (Sec. 6.6).

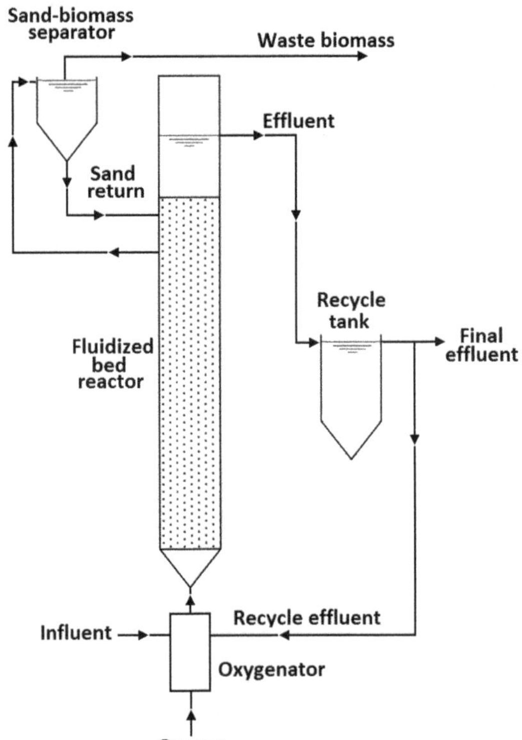


**Fig. 6.2.** The Dorr–Oliver Oxitron™ pilot.

Due to problems of excessive biofilm that leads to high biomass within the liquid phase of the conventional upward flow FBRs (known as upward-fluidization), which affects the proper mixing of the medium with the wastewater, an alternative design has been developed. In comparison to conventional upward fluidization where the medium is denser than water (and so is expanded by the upward flow of wastewater), the media used in downward or inverse flow FBRs (inverse-fluidization) is less dense than water (e.g., polypropylene particles — density of 0.910), which means it is always rising to the surface of the reactor. While inverse fluidized bed reactors (IFBRs), which are three-phase systems, provide better media hydrodynamics within the reactor, it does require a high and superficial fluid velocity to counteract the buoyancy of the medium. Fluidization in IFBRs normally occurs when a downward flow of wastewater meets a counter-current of air (Fig. 6.3), although concurrent wastewater and airflow upward from the base are also used, with the air bubbles causing turbulence within the column, causing the media to disperse a system known as inverse turbulent bed reactors (ITBRs) (Sokól *et al.*, 2009). Internal draft tubes can also create an air-lift effect to ensure good circulation of media and wastewater. IFBRs are cheaper

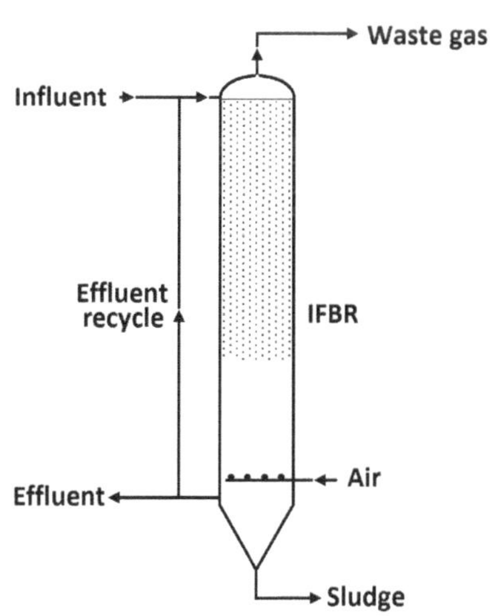


**Fig. 6.3.** Schematic diagram of a typical three-phase inverse fluidized bed reactor (IFBR).

to construct than conventional systems, have high removal efficiencies, are resistant to temperature variations and shock loads, and are ideal for treating low strength wastewaters containing recalcitrant compounds (Swain *et al.*, 2018). IFBRs have been used to treat a wide variety of wastewaters using both aerobic and anaerobic modes such as food processing and brewery wastewaters, phenol wastewater, and acidic, sulfate and metal wastewaters (Bello *et al.*, 2017; Swain *et al.*, 2018).

On-site bioreactors, such as fluidized beds, submerged aerated filters, or biological aerated flooded filters, are widely used for the treatment of industrial and agricultural contaminants in groundwater such as petroleum hydrocarbons, monoaromatic hydrocarbons, chlorinated alipatics and aromatics (Langwaldt and Puhakka, 2000). An unusual application of a fluidized bed to remove algae from Lake Kasumigaura in Japan, a eutrophic lake, is described by Tanaka *et al.* (2001). The system is floated on the lake and powered by solar panels. Three reactors are connected sequentially through the head tanks (Fig. 6.4). Using 0.5 mm diameter granular media, a 50:50 mixture of cristobalite and activated carbon particles, the volume of the media expands about 30% with a hydraulic retention time (HRT) for the entire system of one hour. No process air is injected. Instead, it is the

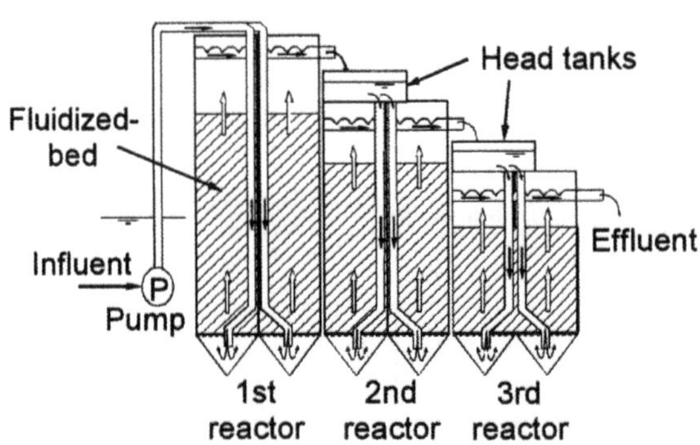


**Fig. 6.4.** Schematic diagram of the experimental fluidised bed biofilm reactor of Tanake *et al.* (2000), which can be used to treat eutrophic lakes. Reproduced with permission of IWA Publishing.

dissolved oxygen in the influent water and that supplied by the cascades in the head tanks mounted on the inlets of the second and third reactors. The system was operated at a flow rate of 1,500 $m^3d^{-1}$ during summer (August to September) when algal blooms occurred; the average efficiency of chlorophyll-*a* removal was 64% at an average influent chlorophyll-*a* concentration of 137.8 $\mu$g $L^{-1}$. Almost all of the algae removed by the reactor were biologically degraded, with 3.47 mg of oxygen required for the aerobic degradation of 1 mg of algal carbon (Kobayashi *et al.*, 1990). Aerobic degradation of algae can be expressed as:

$$C_{106}H_{263}O_{110}N_{16}P + 138O_2 \rightarrow$$
$$106CO_2 + 16N_3^- + HPO_4^{2-} + 122H_2O + 18H^+.$$

Fluidized beds have been extensively studied and reviewed and are considered to be one of the more promising FFR technologies for the future (Bello, *et al.*, 2017; Swain *et al.*, 2018; Özkaya *et al.*, 2019).

## 6.3 Biological Aerated Flooded Filters

### 6.3.1 *Introduction*

Although the correct name for the process is biological aerated flooded filters (BAFFs), this name has been universally shortened to biological aerated filter or biological active filter (BAF) in the United States. Experiments in aerating flooded fixed biofilm reactors have been going on since 1913, although the process we recognize today was first introduced in the late 1980s (Hodkinson *et al.*, 1999). With the introduction of the Urban Wastewater Treatment Directive in Europe, there was pressure to uprate existing sewage treatment plants, especially at coastal sites where land is at a premium, in order to meet new standards. The potential of BAFF technology was immediately recognized because of its compactness, quietness of operation, and the low odor potential of the process. There are many proprietary BAFF systems available, all of which are slightly different in design. However, the principle of the process is common to all (Lilly *et al.*, 1992).

## 6.3.2 *Process Design*

BAFF technology can be used in the same way as the activated sludge process for carbonaceous removal, nitrification, denitrification and biological phosphorus removal (Smith and Hardy, 1992). In fact, whenever activated sludge is appropriate, then BAFFs should also be considered.

Primary sedimentation or fine screening (1–3 mm) is required before secondary treatment by BAFFs. They are constructed above ground in either concrete or steel and must be strong enough to support the media when expanded during backwashing. The system is made up of a number of individual cells or compartments. Mineral or artificial granular medium, generally in the range of 1–5 mm in diameter, is used. Granular media with a density less than water is termed floating (e.g., polyurethane, polystyrene), while media with a density greater than water is termed sunken (e.g., sand, pozzolanic stone). The selection of the medium is a compromise between maximum surface area and the retention efficiency of the medium during backwashing. Like fluidized beds, the granular mineral media in BAFFs are subject to abrasion so that the fine particles formed are lost during backwashing. This leads to media losses of 2–5% per annum. Some proprietary systems employ the fixed lattice type media, more commonly associated with SAFs. The advantage of this modular is that there are lower head loss and no loss of media. However, modular media has significantly lower surface areas, poorer solids removal, and is unable to produce high quality (<10 mg $L^{-1}$ BOD) final effluents. Multi-layer granular media beds, similar to rapid sand filters, are employed where primary sedimentation is omitted (Fig. 6.5). Screened and degritted raw wastewater enters the bottom of the filter and rises through graded layers of media. Solids removal is achieved by gravity flushing, up-flow air and water scouring, and flushing.

The depth of the medium varies from 2–4 m for sunken granular media used in the down-flow mode (now largely replaced by up-flow systems), 3–5 m for up-flow granular media systems, and 4–6 m for fixed laminar media, making all BAFFs highly visible with their tall towers. A separate backwash water holding tank is required to store the final effluent used for backwashing, and another tank for holding

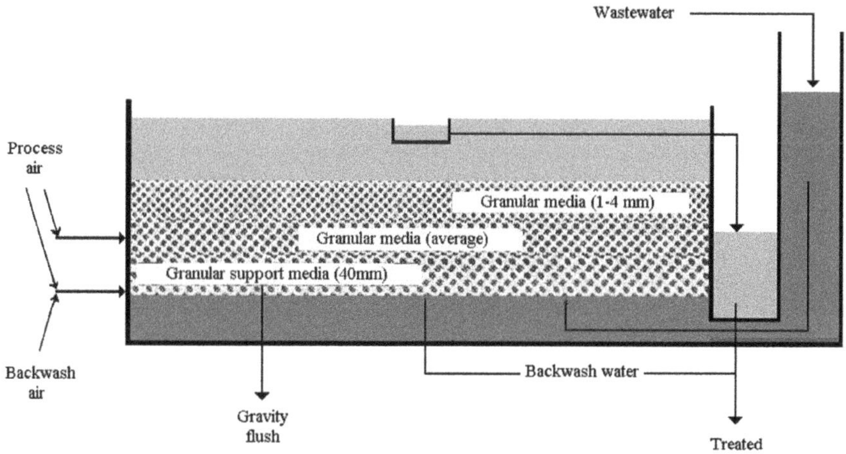


**Fig. 6.5.** A typical multi-layer biological aerated flooded filter (BAFF).

the dirty backwash water to be fed back to the inlet is also required. Air blowers are used for both aeration and air scouring during the backwash cycle. The direction of wastewater flow and backwashing can be in either direction and varies widely (Fig. 6.6).

Two sets of air blowers are required: fine bubble diffusers supply process air at a constant rate to satisfy the oxygen demand of the biomass, while coarse bubble diffusers supply higher rates of air for short periods during backwashing. This often requires two sets of pipework and diffusers and even two sets of blowers. In smaller plants, both fine and coarse bubble aeration can be supplied using perforated latex rubber sheets covered by perforated metal diffuser plates. The size of the bubble and the rate of the air is supplied and controlled by varying the pressure within the pipework, which causes the pores within the latex to expand or contract, thus altering the size of the bubble created.

Backwashing is very important in BAFF systems as it reduces excess biomass from the filter, prevents head loss due to solids retention, and collects the solid for treatment. The frequency of backwashing is controlled by the accumulation rate of the biofilm, which results in head loss through the cell. Typical backwash cycles vary from 12–24 hours for carbonaceous BAFFs to 4–5 days for nitrification-only

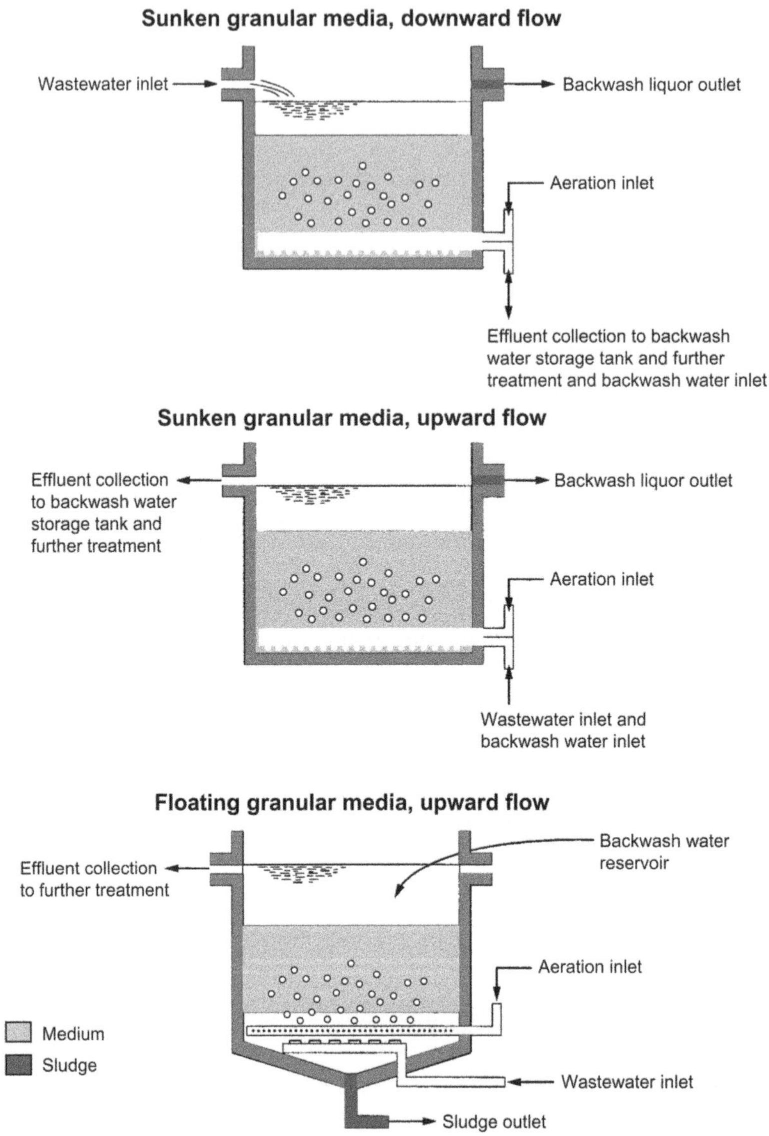


**Fig. 6.6.** Directions of flow and backwash in biological aerated flooded filters (BAFFs).

systems. Backwashing takes between 30–60 minutes and has four phases: (i) Influent feed to the cell is shut off, (ii) water and air are passed upward through the bed of granular medium, expanding it to 1.3–1.5 times its operational depth, (iii) the medium is agitated by air scouring to detach excess biomass, (iv) backwash water is passed upward through the medium to remove the detached biomass. Phases (ii) and (iii) are repeated several times to ensure adequate solids removal. The velocity of backwash water is 5–30 $m^3m^{-2}h^{-1}$ for short periods of 5–20 minutes. The larger the media, the higher the rate. The backwash airflow varies from process to process, but a typical range is 20–60 $m^3$ of air per $m^2$ of bed area each backwash. The volume of backwash water used ranges from 5 to 15% of the average flow, though 10% is typical. It is similar to surplus activated sludge as being a mixture of biomass and trapped solids but with a much lower dry solids content (0.2–0.5%). The backwash water is pumped to the primary sedimentation tank for solids separation or where there is no primary sedimentation, then the backwash water is thickened and the sludge is treated as normal. Where medium losses are high, then the backwash water should be returned up stream of the grit separation stage of the plant.

As BAFFs are truly aquatic ecosystems, the biology is similar to that of activated sludge, although the biomass is largely attached. The biomass is dominated by rotifers, nematodes and the annelid *Aeolosoma hemprichi*. Although the concept of sludge age is inappropriate, the presence of such mesofauna indicates the maturity of the biofilm compared to normal activated sludge biomass. The flooded nature of the system does not favor the development of higher trophic levels, such as invertebrate grazers, which are common in percolating filters, while the backwashing process also physically restricts the size and type of biota found.

## 6.3.3 *Loading and Performance*

Typical organic loadings are between 2–3 kg BOD $m^3d^{-1}$, although this can be increased to 5 kg BOD $m^3d^{-1}$ for high rate treatment producing lower effluent quality (70% BOD). High hydraulic loadings are beneficial because of the improved physical kinetics within the cell, although if the bed becomes fluidized, there will be a significant loss in effluent quality. Assuming one cell is out of operation for backwashing, then hydraulic loadings of 4–12 $m^3m^{-2}h^{-1}$ are normal. Sludge production is between 0.6–1.0 kg of suspended solids per kg BOD removed, while the oxygen requirement of BAFFs is similar to that of activated sludge at 1.0–1.5 kg $O_2$ per kg BOD removed. The design criteria for nitrification are 1.0 kg ammonia — N $m^{-3}d^{-1}$ with an additional oxygen requirement of 4.3 kg $O_2$ kg ammonia — N removed. The current practice is to nitrify in a two-stage BAFF as conditions are more appropriate for nitrifying bacteria in the second stage. An example of a ten-cell steel BAFF unit producing a fully nitrified effluent with an average BOD of 2.7 mg $L^{-1}$ and a suspended solids concentration of 3.7 mg $L^{-1}$ is described by Robinson *et al.* (1994).

Performance falls immediately following backwashing due to a reduction in the physical filtration performance of the medium due to a reduction of the solids that reduce the size of the interstices of the medium. For up to an hour after backwashing, both suspended solids and the BOD will be higher than normal. Nitrification is unaffected by backwashing. Where a 95th percentile effluent quality better than 10 mg $L^{-1}$ BOD and 15 mg $L^{-1}$ suspended solids is required, tertiary treatment is recommended.

Although primarily used for standard secondary treatment, nitrification-denitrification is possible with the BAFF employing similar design criteria as used for nitrifying filters (Meaney and Strickland, 1994). An anoxic zone is required for denitrification with carbon supplied either by recirculation of nitrified effluent or the addition of a separate carbon source (Ryhiner *et al.*, 1994; Ouyang *et al.*, 2000). Volumetric nitrification rates in BAFFs are up to 3.5 and 8 times higher than RBCs and percolating filters using plastic media, respectively (Boller *et al.*, 1997). Tertiary nitrification using BAFFs has also been successfully combined with chemical phosphorus removal. Pujol *et al.* (1995) give an example of a sewage treatment works treating 1,450,000 PE at Köln in Germany. Here BAFFs were used to uprate the existing activated sludge plant for simultaneous nitrification and tertiary phosphorus removal. Removal rates of 70% P [phosphorus] and >95% nitrification were achieved with final effluent concentrations of <0.2 mg N-$NH_4$ $L^{-1}$ and 0.5 mg TP [total phosphorus] $L^{-1}$.

## 6.3.4 *Operational Control*

It takes between 3–8 days for a BAFF system to become operational in terms of BOD removal, although it may take up to 20 days for excellent quality effluents to be achieved. This rapid startup is particularly useful in areas subject to seasonal variation in the population. The high concentration of biomass (15,000–25,000 mg $L^{-1}$) enables high volumetric loading rates compared to the activated sludge process, thereby permitting much smaller tanks to be used.

The operation of BAFF systems is more complex than either activated sludge or percolating filtration due to the frequent backwashing that is required. Therefore, BAFF plants are generally highly automated and operated by computer. Backwashing is operated on a timed basis with an automated head loss override to prevent the cells from overflowing. The quantity of backwash water is significant, and so should be returned to the inlet when the hydraulic loading to the treatment plant is low. In small plants, this is done during the night, while at larger plants, cells are backwashed one at a time continuously. Wherever possible, backwash water should be returned upstream of grit separation to ensure that any displaced medium cannot damage mechanical equipment.

In terms of noise, BAFF plants are similar to a fine bubble diffused air-activated sludge plant. Noise can be reduced even further by using acoustic covers and buildings. Odor potential is low in upward flow BAFF systems but can be significant in downward flow systems where the settled sewage remains on the surface of each cell. The use of diffused air also increases odor potential, but the odor produced from the BAFF unit will be small compared to the odor production of the plant as a whole. Foaming is a common problem in BAFFs, with the foam similar in nature to that found at activated sludge plants. Foaming is normally caused by low loading, so control is achieved by increasing the organic loading by taking some of the cells out of operation. Other control options include recirculation of flow, reduction in the frequency of backwashing, or the use of anti-foaming agents.

The compactness of the BAFF process has led to its widespread use at coastal locations. However, a common problem at such sites is that where the sewerage network is in poor repair, seawater enters the sewer and becomes mixed with the wastewater. Varying concentrations of chloride, or very high concentrations of chloride, cause the bacteria in the BAFF unit to form extra-cellular excretions of polysaccharides, which rapidly block the interstices of granular media. Head loss quickly follows, triggering backwashing prematurely of the timer resulting in short run times and the production of large quantities of backwash water. Operational problems are discussed by Wheale and Cooper-Smith (1995).

## 6.4 Submerged Aerated Filters

Submerged aerated filter (SAF) and BAFF technologies are similar, i.e., the media is submerged and air is supplied via blowers to maintain the biomass on the medium surface, except that SAF systems do not employ backwashing but require a sedimentation tank for solids separation and removal. The medium used is normally modular plastic, 3–6 m in depth, which reduces head loss and prevents blockages (Fig. 6.7). The surface area of the modular medium is significantly lower than the granular media used in BAFFs, which results in lower solids retention that eliminates backwashing. Typical specific surface areas of the SAF modular media varies from 250 to 350 $m^2m^{-3}$; the higher surface areas are used for nitrification. Air is supplied by diffusers, and although only process air is required, some designs incorporate the provision of air scouring at similar pressures used during the backwashing of BAFFs. This provides greater control of sludge removal. Otherwise, the biomass sloughs off in large clumps, causing problems during the final settlement.

Submerged aerated filters are packaged units designed for typical organic loadings of 3 kg BOD $m^{-3}d^{-1}$ and hydraulic loadings of

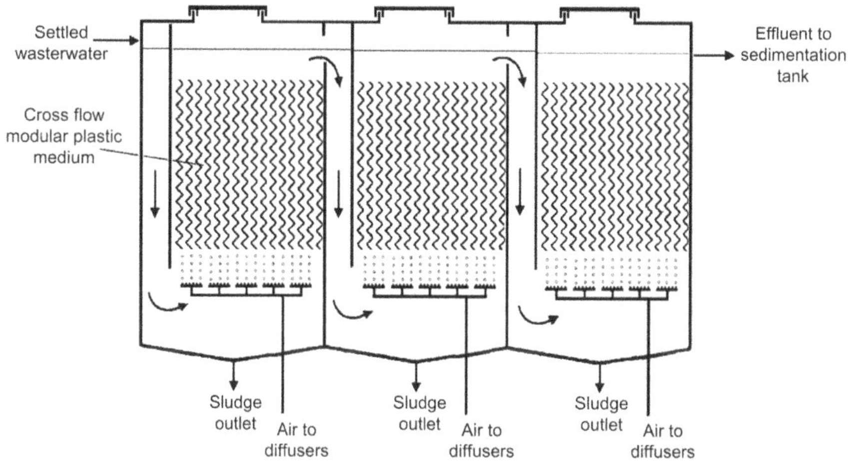


**Fig. 6.7.** Schematic diagram of submerged aerated filters in series.

2–8 $m^3m^{-2}h$. Oxygen requirements and sludge production rates are similar to those for the BAFF. If a 95th percentile effluent quality is <25 mg $L^{-1}$ BOD and 35 mg $L^{-1}$ suspended solids is required, then tertiary treatment must also be used.

Simple to operate and robust, SAFs are ideal packaged units for small treatment plants (<5,000 PE). They require little maintenance, thus making them suitable for unmanned and isolated works. SAF units are used to uprate existing works, for example, by reducing the loads to existing percolating filters in order to promote nitrification. The period for set up is much shorter than for activated sludge, being only 3–8 days for BOD removal, which makes them very flexible in terms of expanding treatment capability. Schlegel and Teichgräber (2000) describe the successful application of SAFs for the pre-treatment of industrial wastewaters, including wastewaters from food processing, carpet dyeing, pharmaceutical and tar processing factories. An upflow SAF system was used by Timur (2001) for denitrification, with maximum denitrification rates achieved at COD:$NO_3$-N ratios of 5–6 that were independent of the hydraulic loading rate or influent $NO_3$-N concentrations. Denitrification followed half-order kinetics, with removal efficiencies of 71–99% reported.

## 6.5 Moving Bed Biofilm Reactor

The principle of the MBBR is small plastic media, normally polyethylene, 10–20 mm in diameter of a density close to that of water (0.90–0.95 g $cm^{-3}$) on which biofilm readily develops (Fig. 6.8). The density of the medium, often referred to as biocarriers, is critical as it allows itself to move freely within the reactor, even when filling. The media is retained within the reactor by a perforated steel barrier, allowing final effluent with any detached solids to pass freely into the secondary settlement tank. Where used on their own, MBBRs do not have sludge return as all the required biomass is retained on the media. Illustrations and details of the commercially available MBBR media are shown in the review by di Biase *et al.* (2019). In contrast to other fixed-film systems, the reactor is completely mixed, so that the whole reactor volume is biologically active, resulting in higher biomass activity (Ødegaard *et al.*, 1994). The system is a hybrid between the activated sludge and a standard submerged fixed-film reactor capitalizing on the advantages of both systems. It has all the advantages of a submerged filter with no clogging problems and a lower head loss (Fig. 6.8). MBBRs have all the flexibility of activated sludge in terms of tank configuration and operation, being operated in either batch or continuous flow modes, in series or parallel, with aerobic, anoxic, anaerobic cycles or tanks, and permitting the full

**Fig. 6.8.** Example of a moving bed biofilm reactor (MBBR) media with biofilm. Reproduced with permission of Headworks Inc., Houston, Texas.

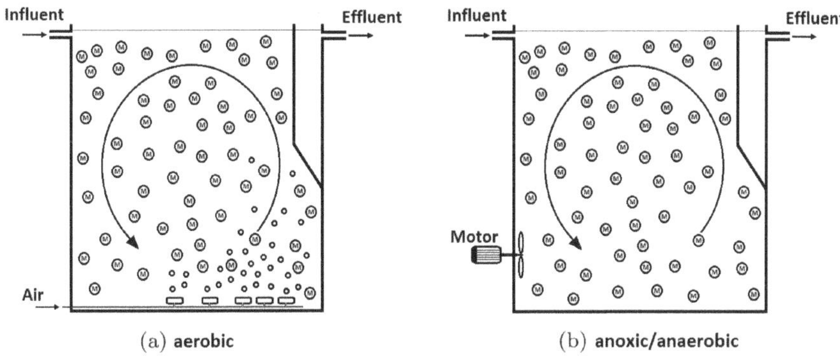


**Fig. 6.9.** Basic reactor designs for moving bed biofilm reactors (MBBRs) with a media retention barrier with (a) aerobic reactors relying on a diffuser for aeration and mixing, and (b) anoxic/anaerobic reactors employing mechanical mixing to maintain a random media in suspension and fully mixed conditions. Adapted from Ødegaard (2006) with permission by IWA Publishing.

range of carbonaceous, nitrification, denitrification and phosphorus removal options (Andreottola *et al.*, 2000). It can be used as an adjunct or upgrade to activated sludge, as a pre-treatment step for stronger wastewaters or a tertiary nitrogen removal step, or as a fully integrated MBBR process incorporating nitrification-denitrification and anammox (Fig. 6.9).

Compared to the activated sludge process, MBBRs can operate with more reactors in series, at higher f/m ratios, and with more selected biomass for each step (Pastorelli *et al.*, 1997; Rusten *et al.*, 1997; Andreottola *et al.*, 2000). Developed in Norway, the first reactor was installed in Lardnal in 1989, specifically to help reduce nitrogen loading in the North Sea. Today, over 400 full-scale MBBRs are found successfully operating throughout the world treating both municipal and industrial wastewaters (e.g., pharmaceutical, textile dye and food processing wastewaters) for BOD/COD removal, nitrification and denitrification (Rusten *et al.*, 1992; Ødegaard *et al.*, 2000; Park *et al.*, 2010; Xing *et al.*, 2014; di Biase *et al.*, 2019). They are also successfully employed in households or small community packaged wastewater treatment units.

Andrettola *et al.* (2000) compared an MBBR system containing Flocor-RMP™ plastic media with a specific surface area of 160 $m^2m^{-3}$

**Table 6.1.** Characteristics of a typical plastic medium (Flocor RMP™) used in moving bed biofilm reactors.

| **Material** | **Polypropylene (density = 0.94 g cm³)** | |
|---|---|---|
| Shape | Corrugated cylinder | |
| Dimensions | Length: | 20–30 mm |
| | Diameter: | 15–20 mm |
| Specific surface | 160 $m^2/m^3$ | |
| Filling rate | 70% | |

(Table 6.1) with activated sludge. As only 70% of the tank was occupied by the medium, the actual active medium surface area, termed the specific MBBR surface, was 112 $m^2m^{-3}$. The performance of the two processes was almost identical, and they concluded that media with higher surface areas were needed to achieve better performance by increasing the biomass. However, MBBRs require no sludge recycling, do not require close control as with activated sludge, nor has the associated bulking problems. Specialized biomass for carbon and nitrogen removal can be selected by using multi-reactor configurations.

Ødegaard *et al.* (2000) have analyzed the influence of media size and shape on performance. They found that MBBRs should be designed based on surface area loading rates (g COD $m^2d^{-1}$), while the shape and size of the media is not a significant factor as long as an effective surface area is provided. Design values as high as 30 mg soluble COD $m^2d^{-1}$ can be used. However, problems were reported that the settleability of the biomass decreased with increasing organic loading. MBBR technology has been reviewed by Ødegaard (2006) and di Biase *et al.* (2019).

## 6.6 Integrated Fixed-Film Activated Sludge Systems

IFAS systems differ from MBBR reactors in that the activated sludge is returned to the reactor so that the activated sludge floc and biofilm on the media are of equal importance in terms of treatment.

However, they overlap in terms of design and can be seen as a variation of the MBBR process rather than a process in its own right. The IFAS is used either as a standalone design or a simple method of increasing the loading to conventional diffuse air activated sludge reactors, the key benefit being enhanced performance without having to increase the MLSS of the activated sludge, thereby not increasing the solid loading on the secondary settlement tanks, although care must be taken not to increase the hydraulic loading beyond their design limit. The reason why it is a separate process is that originally activated sludge reactors were upgraded using fixed modular media or conventional random percolating (trickling) filter media held in cages placed within the reactor to prevent carryover of the medium into the final settlement phase. This type of media and its placement in cages led to problems of blockages within the media, leading to anaerobic conditions that required more sophisticated media and housing. There are now specially designed IFAS modular media, mainly of the rope or mesh type (e.g., Ringlace™, Bioweb™ and Accweb™), which prevents fouling by excessive biofilm development, along with standard random MBBRs or very similar media specifically for use in IFAS systems with specific surface areas of between 350–1,000 $m^2m^{-3}$ (Fig. 6.10). As in MBBR systems, a porous steel barrier is used to retain the media in the reactor while allowing the mixed liquor to pass into the secondary settlement tank where the sludge is separated and returned with media-filling fractions varying up to 50–60%. The IFAS process can be used in many different combinations resulting in many different modifications, and like MBBRs, the process is extremely flexible and able to achieve not only carbonaceous treatment but also nitrification, denitrification, anammox and biological phosphorous removal. The presence of the media biofilm in the activated sludge tank creates two different microbial communities, each with a very different sludge residence time (SRT). This is particularly useful for nitrification, as nitrifying bacteria require significantly longer SRTs (>15 days) compared to the normal heterotrophic bacteria that dominate the suspended flocs, including polyphosphate-accumulating bacteria that prefer shorter SRTs (Onnis-Hayden *et al.*, 2011).

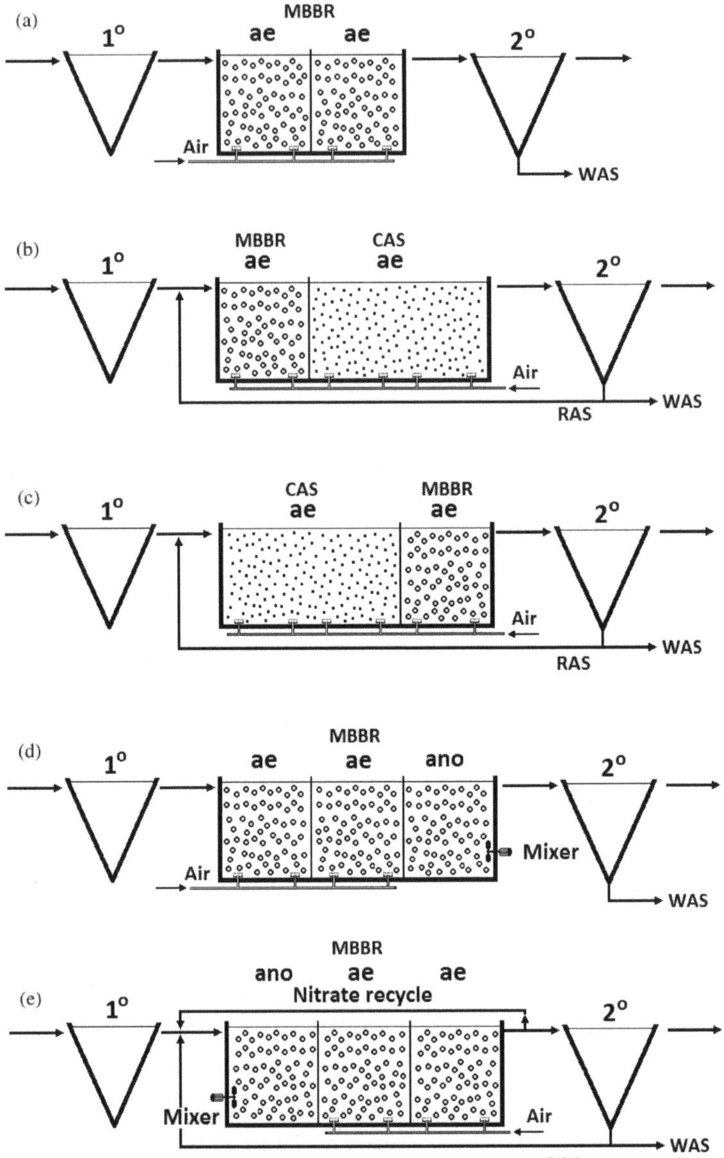


**Fig. 6.10.** Common configurations of moving bed biofilm reactors (MBBRs) where **ae** is an aerobic zone and **ano** is an anoxic zone, with a primary (**1°**) and secondary (**2°**) settlement. (a) Standard MBBR without the recycling of waste sludge (WAS), (b) MBBR pre-treatment for upgrading conventional activated sludge (CAS) plants with normal sludge return (RAS), (c) MBBR addition to CAS for nitrification, (d) post denitrification, and (e) pre-denitrification with RAS being optional for hybrid reactors.

IFAS processes are varied in design and are currently primarily used for decentralized wastewater treatment (DEWATS) (Singh and Kazmi, 2016), upgrading the loading to existing activated sludge, and for additional nitrogen removal. However, the IFAS process has also been successfully used to treat complex industrial wastewaters, including trace organics (Shreve and Brennan, 2019). The early processes utilized reticulated polyurethane foam pads, which were added to the main reactor. Reticulated foam is very different from normal foams in that the windows between the cells have been chemically removed, thus giving a very open structured and highly porous (97%) medium. Different media can have pore sizes of 300–2,500 μm. The structure also allows larger organisms, such as nematodes and oligochaetes, to colonize the pads.

One of the first commercial processes was CAPTOR™, which uses rectangular pads (25 × 25 × 12 mm) with a nominal pore size of 850 μm (Sec. 6.2). Between 20,000 and 50,000 pads are added per cubic meter of the reactor, which takes up between 15 and 75% of the aeration tank volume. This allows very high biomass (MLSS) concentrations of 7,000 to 9,000 mg $L^{-1}$, which is equivalent to 100–150 mg MLSS per pad. The pads are continuously removed and mechanically cleaned by compression, using a mini sludge belt press producing a waste sludge at 6% of dry solids, the fall back into the reactor. In practice, the pads are difficult and messy to clean, and when new or desludged pads are returned to the aeration tank, they often float rather than re-enter suspension. However, the process has been used for a variety of applications, including nitrification and denitrification (Cooper, 1989). The LINPOR™ process is similar to CAPTOR™, although much smaller pads are used (12 × 12 × 12 mm), and a secondary settlement tank is required. Biological growth on the pads is controlled by shear and so can be retained in the reactor permanently, with the suspended solids, including the sheared biomass from the media, which separated from the final effluent during secondary settlement. Only 10 to 40% of the tank volume was taken up by the medium, with the biomass concentrations (5,000–8,000 mg MLSS $L^{-1}$) double of that found in conventional systems activated sludge systems. The process is used for combined carbonaceous and

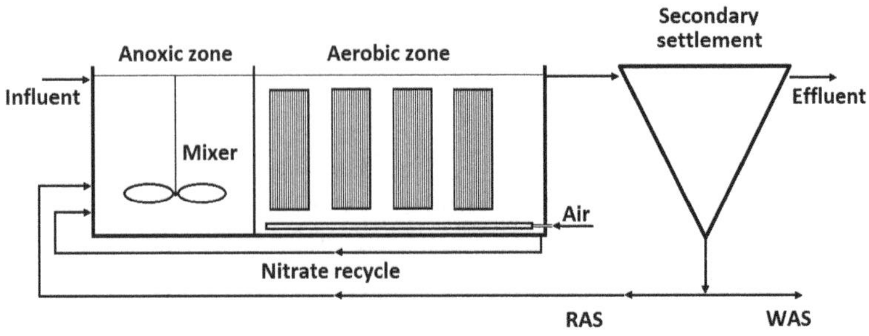


**Fig. 6.11.** Schematic of a commonly employed IFAS design with pre-denitrification using modular units of rope media.

nitrification (LINPOR-CN™) or as a tertiary nitrification-only process (LINPOR-N™). However, neither process has been widely adopted commercially (Morper, 1994).

Cleartec® is a typical example of the modular type media now available and is made up of layers or curtains of hanging curlz, a very strong loop knotted polypropylene fabric that is housed in rigid boxes that are lowered into the aeration tank. The biomass grows over the loose threads, forming a dense but highly porous biofilm structure that allows diffuse air to pass freely through (Fig. 6.11). Cleartec® is different in that the material promotes a thinner and more uniform biofilm, especially suited for promoting nitrification. This type of media offers a very simple retrofit option to overloaded activated sludge systems, and also offers significant denitrification capacity within nitrifying systems as the nitrate diffuses deeper into the biomass where anoxic conditions will be present. In their experimental single-stage MBBR using Cleartec® media, Singh and Kazmi (2016) treated municipal wastewater and recorded mean removal rates of 92% COD, 91% BOD, 90% TSS [total suspended solids], 88% TN [total nitrogen] and 50% TP with a 2.2 and 2.4 log reductions in total and fecal coliforms, respectively. The system was operated with a dissolved concentration of 3 mgL$^{-1}$, HRT of 6.9 hours, and a sludge age of 7 days. They found that the attached biomass was more porous and homogenous compared to the suspended activated sludge flocs, with twice the total biomass supported on the media than suspended. Not surprisingly, their respirometric evaluation showed that the majority of the activity and hence treatment was occurring within the attached biofilm.

Dissolved oxygen (DO) concentrations in IFAS reactors need to be able to support the high biomass associated with the media as well as the MLSS and can affect performance significantly. Although DO concentrations for IFAS reactors are usually given at between 2–3 mg $L^{-1}$, Singh *et al.* (2016), through a series of experiments on a single-stage pilot IFAS reactor, found optimal DO concentrations of 4.5, 4.5 and 2.5 mg $L^{-1}$ for carbonaceous removal, nitrification and denitrification, respectively. These values are considerably higher than for activated sludge alone, thus adding significantly to energy costs. They also observed that the ciliate protozoan population found in the reactors reflected the oxygen concentration.

In a direct comparison of the IFAS system with the MBBR for nitrogen removal from municipal wastewater by partial nitritation/anammox, Malovanyy *et al.* (2015) found that the IFAS system was far superior in terms of efficiency and removal rates.

The advantages of the IFAS system over activated sludge-only systems are: (i) better ability to deal with shock loadings, (ii) increased biological activity per unit volume of the reactor, (iii) no increase in solids loading to the final sedimentation tanks, (iv) improved sludge settleability shown by a reduced sludge volume index, (v) lower sludge production, (vi) reduced footprint per kg BOD removed, (vii) flexibility in treatment capacity by adding/removing media, and (viii) simultaneous nitrification/denitrification possible. Disadvantages are restricted to media handling and a higher level of operational management is required. Examples of IFAS plants are given by WEF (2011).

## References

Ahmadi, M., Alian, A. M., Amiri, H., Ramavandi, B. and Izanloo, H. (2014) Upgrading of Kish Island Mirmohana wastewater treatment plant using moving bed biofilm reactor alternative. *Iranian Journal of Health Sciences*, **2**, 33–42.

Andreottola, G., Foladori, P., Ragazzi, M. and Tatano, F. (2000) Experimental comparison between MBBR and activated sludge system for the treatment of municipal wastewater. *Water Science and Technology*, **41**, (4/5), 375–382.

Bello, M.M., Raman, A.A.A. and Purushothaman, M. (2017) Applications of fluidized bed reactors in wastewater treatment — A review of the major design and operational parameters. *Journal of Cleaner Production*, **141**, 1492–1514.

Boller, M., Tschui, M. and Gujer, W. (1997) Effects of transient nutrient concentrations in tertiary biofilm reactors. *Water Science and Technology*, **36**, (1), 101–109.

Cooper, P. (1989) Demonstration and evaluation of the CAPTOR process for sewage treatment. EPA-600/S2-88-060. US Environmental Protection Agency Cincinnati, Ohio, USA.

Cooper, P.F. and Wheeldon, D.H.V. (1980) Fluidised and expanded bed reactors for wastewater treatment. *Water Pollution Control*, **79**, 286–306.

Cooper, P.F. and Wheeldon, D.H.V. (1982) Complete treatment of sewage in a two-stage fluidised-bed system. *Water Pollution Control*, **81**, 447–464.

di Biase, A., Kowalski, M.S., Devlin, T.R. and Oleszkiewicz, J.A. (2019) Moving bed biofilm reactor technology in municipal wastewater treatment: A review. *Journal of Environmental Management*, **247**, 849–866.

Gauntlett, R.B. and Craft, D.G. (1979) Biological removal of nitrate from river water. Technical Report: TR98. Water Research Centre, Stevenage, UK.

Hodkinson, B., Williams, J.B. and Butler, J.E. (1999) Development of biological aerated filters: A review. *Journal of the Chartered Institution of Water and Environmental Management*, **13**, (4), 250–254.

Hoyland, G. and Robinson, P.J. (1983) Aerobic treatment in "Oxitron" BFB plant at Coleshill. *Water Pollution Control*, **82**, 479–493.

Jeris, J.S., Owens, R.W., Hickey, R. and Flood, F. (1977) Biological fluidized-bed treatment for BOD and nitrogen removal. *Journal of the Water Pollution Control Federation*, **49**, 816–831.

Kobayashi, S., Uno, K. and Yoshizawa, T. (1990) Characteristics of COD, nitrogen, and phosphorus in water of Lake Inba Numa and Lake Teganuma — COD originated in phytoplankton and transformation rate of nitrogen and phosphorus into COD. *Kogai to Taisaku*, **26**, (14), 47–56.

Langwaldt, J.H. and Puhakka, J.A. (2000) On-site biological remediation of contaminated groundwater: A review. *Environmental Pollution*, **107**, (2), 187–197.

Lilly, W., Bourn, G., Crabtree, H., Upton, J. and Thomas, V. (1992) The production of high quality effuents in sewage treatment using the bicarbone process. *Journal of the Institution of Water and Environmental Management*, **6**, (2), 179–184.

Malovanyy, A., Trela, J. and Plaza, E. (2015) Mainstream wastewater treatment in integrated fixed film activated sludge (IFAS) reactor by partial nitritation/anammox process. *Bioresource Technology*, **198**, 478–487.

Meaney, B.J. and Strickland, J.E.T. (1994) Operating experiences with submerged filters for nitrification and denitrification. *Water Science and Technology*, **29**, (10–11), 119–125.

Morper, M.R. (1994) Upgrading of activated sludge systems for nitrogen removal by application of the LINPOR®-CN process. *Water Science and Technology*, **29**, (12), 167–176.

Moteleb, M.A., Suidan, M.T., Kim, J., Davel, J.L. and Adrian, N.R. (2001) Anaerobic degradation of 2,4,6-trinitrotoluene in granular activated carbon fluidized bed and batch reactors. *Water Science and Technology*, **43**, (1), 67–75.

Ødegaard, H. (2006) Innovations in wastewater treatment: The moving bed biofilm reactor process. *Water Science and Technology*, **53**, (9), 17–33. https://doi.org/10.2166/wst.2006.284.

Ødegaard, H., Gisvold, B. and Stickland, J. (2000) The influence of carrier size and shape in the moving bed biofilm process. *Water Science and Technology*, **41**, (4–5), 383–392.

Ødegaard, H., Rusten, B. and Westrum, T. (1994) A new moving bed biofilm reactor — Applications and results. *Water Science and Technology*, **29**, (10–11), 157–165.

Onnis-Hayden, A., Majed, N., Schramm, A. and Gu, A.Z. (2011) Process optimization by decoupled control of key microbial populations: Distribution of activity and abundance of polyphosphate-accumulating organisms and nitrifying populations in a full-scale IFAS-EBPR plant. *Water Research*, **45**, (13), 3845–3854.

Ouyang, C.F., Chiou, R.J. and Lin, C.T. (2000) The characteristics of nitrogen removal by the biofilm system. *Water Science and Technology*, **42**, (12), 137–147.

Özkaya, B., Haksonen, A.H., Sahinkaya, E. and Puchakka, J.A. (2019) Fluidized bed bioreactor for multiple engineering solutions. *Water Research*, **150**, 452–465.

Park, H.O., Oh, S., Bade, R. and Shin, W.S. (2010) Application of A20 moving-bed biofilm reactors for textile dyeing wastewater treatment. *Korean Journal of Chemical Engineering*, **27**, 893–899.

Pastorelli, G., Andreottola, G., Canziani, R., Darriulat, C., de Fraja Frangipane, E. and Rozzi, A. (1997) Organic carbon and nitrogen removal in moving bed biofilm reactors. *Water Science and Technology*, **35**, (6), 91–99.

Pujol, R., Peladen, J.G. and Buchl, K.H. (1995) Nutrient removal by up-flow filtration. *Proceedings of the 7th Conference on Design and Operation of Large Wastewater Treatment Plants*, Vienna, 27 August to 1 September, pp. 245–253.

Robinson, A.B., Brignal, W.J. and Smith, A.J. (1994) Construction and operation of a submerged aerated filter sewage treatment works. *Journal of the Institution of Water and Environmental Management*, **8**, (2), 215–221.

Rusten, B., Kolkinn, O. and Ødegaard, H. (1997) Moving bed biofilm reactors and chemical precipitation for high effciency treatment of wastewater for small communities. *Water Science and Technology*, **35**, (6), 71–79.

Rusten, B., Ødegaard, H. and Lundar, A. (1992). Treatment of dairy wastewater in a novel moving bed biofilm reactor. *Water Science and Technology*, **26**, 703–711.

Ryhiner, G., Sorensen, K., Birou, B. and Gros, H. (1994) Biofilm reactors configured for advanced nutrient removal. *Water Science and Technology*, **29**, (10/11), 111–125.

Schlegal, S. and Teichgraber, B. (2000) Operational results and experience with submerged fixed film reactors in the pre-treatment of industrial effuents. *Water Science and Technology*, **41**, (4/5), 453–459.

Shreve, M.J. and Brennan, R.A. (2019) Trace organic contaminant removal in six full-scale integrated fixed-film activated sludge (IFAS) systems treating municipal wastewater. *Water Research*, **151**, 318–331.

Singh, N.K. and Kazmi, A.A. (2016) Environmental performance and microbial investigation of a single stage aerobic integrated fixed-film activated sludge (IFAS) reactor treating municipal sewage. *Journal of Environmental Chemical Engineering*, **4**, 2225–2237.

Singh, N.K., Kazmi, A.A. and Starkl, M. (2016) Treatment performance and microbial diversity under dissolved oxygen stress conditions: Insights from a single stage IFAS reactor treating municipal wastewater. *Journal of the Taiwan Institute of Chemical Engineers*, **65**, 197–203.

Smith, A.J. and Hardy, P.J. (1992) High-rate sewage treatment using biological aerated filters. *Journal of the Institution of Water and Environmental Management*, **6**, (2), 179.

Sokól, W., Ambaw, A. and Woldeyes, B. (2009) Biological wastewater treatment in the inverse fluidised bed reactor. *Chemical Engineering Journal*, **150**, 63–68.

Sutton, P.M., Shleh, W.W., Kos, P. and Dunning, P.R. (1981) Dorr-Oliver's "Oxitron" fluidized bed water and wastewater treatment process. In *Biological Fluidised Bed Treatment of Water and Wastewater* (eds. P.F. Cooper and B. Atkinson). Ellis-Horwood, Chichester, UK, pp. 285–304.

Swain, A.K., Sahoo, A., Jena, H.M. and Patra, H. (2018) Industrial wastewater treatment by aerobic inverse fluidized bed biofilm reactors (AIFBBRs): A review. *Journal of Water Process Engineering*, **23**, 61–74.

Tanaka, T., Suzuki, K., Nishijima, N. and Takagi, T. (2001) Algae-removal performance of a fluidized-bed biofilm reactor system for lake water treatment. *Water Science and Technology*, **43**, (1), 277–283.

Timur, H. (2001) Denitrification of wastewater in a fixed film reactor. *Fresenius Environmental Bulletin*, **10**, (4), 396–400.

van Haandel, A.C. and van der Lubbe, J.G.M. (2012) *Handbook of Biological Wastewater Treatment, 2nd ed.* IWA Publishing, New York.

Walker, I. and Austin, E.P. (1981) The use of plastic porous biomass supports in a pseudo-fluidised bed for effuent treatment. In *Biological Fluidised Bed Treatment of Water and Wastewater* (eds. P.F. Cooper and B. Atkinson). Ellis-Horwood, Chichester, UK, pp. 272–284.

WEF (2011) Biofim reactors. WEF Manual of Practice No. 35. Water Environment Federation Press, Alexandria, Virginia, US.

Wheale, G. and Cooper-Smith, G.D. (1995) Operational experience with biological aerated filters. *Journal of the Chartered Institution of Water and Environmental Management*, **9**, (2), 109–118.

Xing, Z.P., Sun, D.Z., Yu, X.J., Zou, J.L. and Zhou, W. (2014) Treatment of antibiotic fermentation-based pharmaceutical wastewater using anaerobic and aerobic moving bed biofilm reactors combined with ozone/hydrogen peroxide process. *Environmental Progress and Sustainable Energy*, **33**, 170–177.

# 7

# Nitrogen Removal and Filters

## 7.1 The Processes

By the time wastewater reaches the treatment plant, the majority of the nitrogen present is in the form of ammonium ($NH_4$). Concentrations in wastewaters vary from <5 mg N $L^{-1}$ for domestic sewage, and <100 mg N $L^{-1}$ in municipal wastewater for up to 5,000 mg N $L^{-1}$ in industrial wastewaters. Concentrations exceeding this are treated by physico-chemical methods, normally air-stripping.

Fixed-film reactors (FFRs) are involved in a number of nitrogen oxidation and reduction processes:

| Process | Oxidation/Reduction Reaction |
|---|---|
| Nitritation | Ammonium ($NH_4^+$) oxidized to nitrite ($NO_2^-$) |
| Nitratation | Nitrite ($NO_2^-$) oxidized to nitrate ($NO_3^-$) |
| Nitrification | Ammonium ($NH_4^+$) and nitrite ($NO_2^-$) oxidized to nitrate ($NO_3^-$) |
| Denitritation | Nitrite ($NO_2^-$) reduced to dinitrogen gas ($N_2$) |
| Denitrification | Nitrate ($NO_3^-$) reduced to dinitrogen gas ($N_2$) |
| Anammox | Nitrite ($NO_2^-$) and ammonium reduced to dinitrogen gas ($N_2$) |

Where anammox is an acronym for anaerobic ammonium oxidation.

Most FFRs used for domestic treatment achieve full nitrification, this being dependent on the ratio of organic substrate to nitrogen loading, expressed as g COD/g N (i.e., the COD/N ratio). Where this is >20, the need for N by the heterotrophic bacteria for biomass synthesis during carbonaceous removal utilizes most of the N present. As this ratio falls toward 5, then nitrification and denitrification processes are required to remove the excess N not required for heterotrophic bacteria that would otherwise end up in the final effluent leading to pollution. However, as the ratio falls below 5, then insufficient organic substrate limits the degree to which standard heterotrophic denitrification can be achieved without the addition of an extra carbon source. In these cases, partial-nitrification-anammox (PNA) or nitritation-denitritation processes are employed. By using nitrite as the starting point for nitrogen removal, less organic substrate is required, if at all. However, in practice, the anaerobic ammonium oxidizing bacteria are very slow-growing, and so are only effective at temperatures above 25°C, limiting their current use to industrial applications (e.g., wastewaters from pharmaceutical, food processing, distilling, tanning and the semiconductor industries) and the treatment of the wastewater from sludge digesters (Ekama and Wenzel, 2008).

### 7.1.1 *Nitrification*

Nitrification consists of two separate microbial steps carried out by different groups of chemolithoautrophs. Nitritation is carried out by ammonium-oxidizing aerobic bacteria (e.g., *Nitrosomonas* spp.), which also includes some heterotrophs and Archea, while nitratation is carried out by nitrite-oxidizing aerobic bacteria (e.g., *Nitrobacter* spp.):

*Nitritation*

$$NH_4^+ + 1.5O_2 \rightarrow NO_2^- + H_2O + 2H^+$$

*Nitratation*

$$NO_2^- + 0.5O_2 \rightarrow NO_3^-$$

As shown below in the equations, nitrification by chemolithoautotrophic bacteria results in the loss of carbonate ions, thus reducing the alkalinity of the wastewater being treated and possibly lowering the pH. While there is normally sufficient buffering capacity in domestic and municipal wastewaters, it may be a problem in very soft water supply areas or when treating industrial wastewaters, as pH adjustment may be required (Ekama and Wenzel, 2008). In the first reaction, the bacteria use ammonium for both energy and as a nitrogen source, while in the second reaction, the bacteria use nitrite as the energy source and ammonium as a source of nitrogen. Both are aerobic microbial processes using inorganic carbon as a source of carbon.

*Nitritation*

$$NH_4^+ + 1.383O_2 + 0.09HCO_3^- \rightarrow 0.987NO_2^- + 1.036H_2O + 0.018C_5H_7O_2N + 1.892H^+.$$

*Nitratation*

$$NO_2^- + 0.003NH_4^+ + 0.485O_2 + 0.015HCO_3^- + 0.012H^+ \rightarrow NO_3^- + 0.009H_2O + 0.003C_5H_7O_2N.$$

### 7.1.2 *Denitrification*

While a wide range of bacteria and archaea can denitrify under anoxic conditions by using $NO_3^-$ and $NO_2^-$ as terminal electron acceptors, it is generally facultative heterotrophic bacteria that are the major driver in wastewater treatment. Reduction of nitrate follows a sequence of steps, each mediated by a specific enzyme.

$$NO_3^-(aq) \rightarrow NO_2^-(aq) \rightarrow NO(g) \rightarrow N_2O(g) \rightarrow N_2(g).$$

It is during incomplete denitrification that both NO and $N_2O$ can be released into the atmosphere, causing damaging effects to the environment in terms of toxicity and global warming. The process requires an organic substrate, and so has the effect of reducing the biochemical oxygen demand (BOD) or chemical oxygen demand (COD) of the final effluent. However, where insufficient organic matter is available, then denitrification can only proceed with the addition of an extra carbon source such as methanol or acetic acid.

*Denitrification*

$$NO_3^- + 1.25CH_2O \rightarrow 0.5N_2 + OH^- + 0.75H_2 + 1.25CO_2.$$

Denitrification releases $OH^-$ ions into the effluent, thus increasing the alkalinity.

### 7.1.3 *Anaerobic Ammonium Oxidation (Anammox)*

Although both nitrification and denitrification were observed in the 19th century, anammox was not discovered until the 1990s. The autotrophic bacteria responsible for the oxidation of ammonia and the simultaneous reduction of nitrite belong to the widely distributed phylum *Planctomycetes.* Not all the bacteria in this phylum can oxidize ammonium, this being restricted to the Anammox-*Planctomycetes*, which have a characteristic membrane-bound cell compartment called the anammoxosome (Fuerst and Sagulenko, 2011). Ammonium is used as an energy source by the anammox bacteria and carbon dioxide ($CO_2$) as a carbon source, with nitrite $(NO_2^-)$ and nitrate $(NO_3^-)$, if present, used as electron acceptors. Kartel *et al.* (2011) have described the process in detail, where the enzyme nitrite oxidoreductase reduces nitrite to nitrogen oxide (NO), which then reacts with ammonium to form hydrazine ($N_2H_2$), enabled by the enzyme hydrazine synthase, which is oxidized by hydrazine dehydrogenase into nitrogen gas ($N_2$). The bacteria are extremely slow-growing and strict anaerobes, and are unable to function if the dissolved oxygen concentration rises above 0.5 mg $L^{-1}$. An example of the identification of aerobic ammonium oxidizing and anammox bacteria found in the biofilm from a rotating drum biological contactor is given in Fig. 7.1.

*Anammox*

$$NH_4^+ + NO_2^- \rightarrow N_2 + 2H_2O.$$

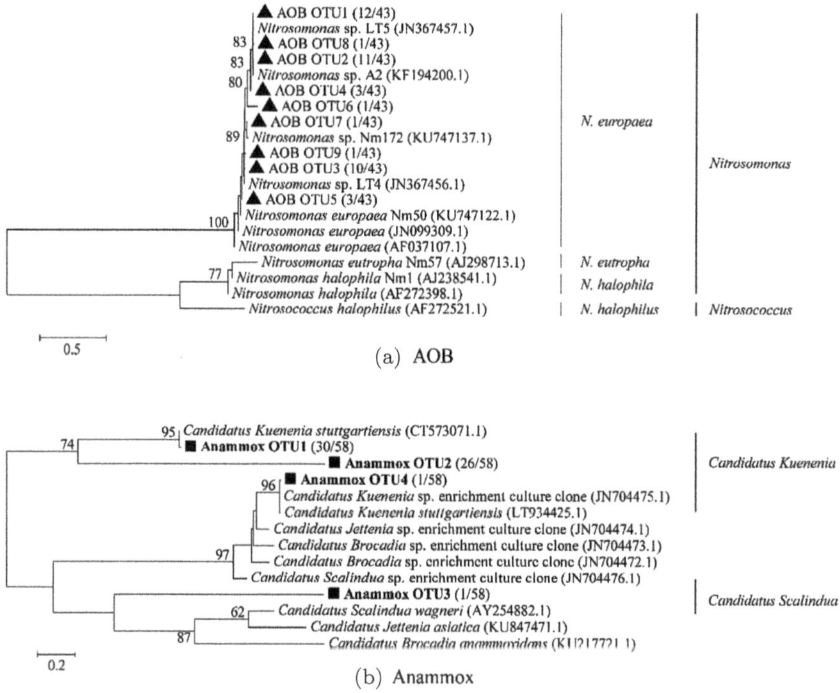


**Fig. 7.1.** Example of a DNA analysis of a nitrifying-denitrifying biofilm taken from a rotating drum biological contactor. Neighbor-joining phylogenetic tree for (a) amoA gene and (b) hzsB gene sequences. Each operational taxonomic unit (OTU) in this study was named AOB OTU# or ANAMMOX OTU#, and the first number in the following parentheses was the number of clones belonging to this OTU, and the second number was the retrieved clones in this sample (Li *et al.*, 2019). Reproduced with permission of Elsevier Publishing.

To facilitate the anammox process, nitrite must be present and can be produced by a separate nitritation process known as a partial nitritation process (PN). Most commercial anammox systems employ PN as an integrated part of the main process (Lackner *et al.*, 2014), including when used as a tertiary step for nitrate removal from domestic wastewater treatment plants (Cao *et al.*, 2019).

## 7.2 Applications

Nitrification and denitrification are an integral part of all standard biological wastewater treatment process designs, and this has already been discussed in the different FFR chapters. However, there are a number of dedicated FFR nitrogen removal processes in use.

### 7.2.1 *Nitrification*

Many biological treatment processes, because of design or loading, can complete the carbonaceous oxidation of wastewater but are unable to successfully oxidize the ammonia-nitrogen ($NH_3$-N) present. In the activated sludge process, failure to provide full nitrification of wastewater may be due to inadequate retention time, sludge age or aeration capacity. In percolating filtration, modifications such as alternating double filtration or high-rate filtration using modular media, insufficient surface area, or high organic loadings may result in incomplete oxidation of ammonia to nitrate. Ammonia-nitrogen is very toxic, and many discharge licenses will include an upper limit for ammonia being discharged to freshwaters. Therefore, a separate purification stage just to remove ammonia is often required to meet such standards.

All FFRs can support nitrification as part of the overall treatment of wastewater, although success depends on a number of factors, in particular, organic loading, sufficient retention time, inhibition of growth of nitrifiers by metals, appropriate C/N ratio in the nitrifying zone, and most importantly temperature. Percolating filters, rotating biological contactors (RBCs), biological aerated filters and fluidized beds have all been specifically used to nitrify effluents after treatment by other biological systems, often followed by dedicated denitrification units or zones.

The sole purpose of nitrifying filters is to remove ammonia from effluents that have undergone complete carbonaceous oxidation. It is based on the stratification of microbial species in conventional low-rate filters, where nitrifying bacteria are located toward the base of the bed. In this way, separate filters were developed for carbonaceous oxidation and nitrification, although nitrifying filters can also be used to nitrify the effluent from other types of treatment units. Nitrifying filters are very similar to percolating filters in design, except that smaller grades of media are used (<25 mm) to maximize the specific surface area (Lekang and Kleppe, 2000). The use of smaller grades of media is made possible as the wastewater contains very little organic matter, resulting in minimal development of the heterotrophic biofilm. Therefore, excessive biofilm accumulation that could cause clogging does not occur. Media selection for nitrifying filters is a compromise between specific surface area, voidage, and the flow characteristics induced by the medium (i.e., crossflow, vertical flow or random flow). The nitrifying bacteria themselves are very slow-growing and produce only a small volume of biomass in comparison with heterotrophs. For example, the cell yield of *Nitrosomonas* spp. is just 0.053 g per g $NH_3$-N oxidized. The film accumulation is also restricted by a high hydraulic loading of between 2–6 $m^3m^{-3}d^{-1}$. Wastewater is distributed on to the filter via a fixed distribution system that uses fine spray nozzles to ensure maximum utilization of the media and adequate oxygen. Using spray nozzles also allows the maximum loss of ammonia by volatization. Removal rates of the ammonia-nitrogen can rise to 120 mg $m^{-3}d^{-1}$, although this depends on the specific surface area of the media, so the removal rates of ammonia-nitrogen per unit surface area are in the order of 500–1,000 mg $m^{-2}d^{-1}$ (Barnes and Bliss, 1983). Wik (1999) has studied strategies to improve the efficiency of nitrifying filters.

Nitrifying filters were widely introduced to control eutrophication in surface waters to comply with increasingly stringent legislation during the 1990s as a tertiary step often to activated sludge systems (Boller and Gujer, 1986). However, they have been largely replaced by upgrading the activated sludge phase to a biological nutrient removal system. Persson *et al.* (2002) describe the use of such filters — percolating filters and RBCs — in Sweden. They found that ammonium oxidizing bacteria (AOB) and biofilm accumulation decrease with depth, with nitrite oxidizing bacteria (NOB) being more restricted at lower depths, and could be reliably modeled. In FFRs, AOB mainly belong to the genera *Nitrosomonas* and *Nitrosospira*, with *Nitrospira* the commonest genera of NOB and *Nitobacter* also present (Burell *et al.*, 1998; Wagner and Amman, 1998).

Nitrifying trickling or percolating filters are widely used to oxidize ammonia from closed-circuit aquaculture farms, where a range of small plastic and expanded clay aggregate media (e.g., Leca), which can be treated with special surface coatings, are used (Lekang and Kleppe, 2000; Gutierrez-Wing and Malone, 2006; Malone and Pfiffer, 2006). Three such media types (e.g., Biofill™, Bactoballs™ and MECHpro™ rings) were compared by Godoy-Olmos *et al.* (2016). Removal rates of up to 0.27 g of total ammonia nitrogen (TAN) $m^2d^{-1}$ were achieved, although this was dependent on the temperature and TAN loading rate.

Performance evaluation and modeling are often complicated in nitrifying filters due to ammonium adsorption onto the biofilm and denitrification at high ammonium loadings (Wik, 1999). The design and operation of nitrifying trickling filters are reviewed by Eding *et al.* (2006).

### 7.2.2 *Denitrification*

There are three possibilities for the inclusion of denitrification in FFRs: Pre-denitrification, post-denitrification and combined nitrification-denitrification. Some FFR systems such as moving bed biofilm reactors (MBBRs) or integrated fixed-film activated sludge (IFAS) systems are particularly well suited for combined nitrification-denitrification (Secs. 6.5 and 6.6).

Nitrifying filters are often followed by submerged denitrifying filters, especially when treating high strength ammonium wastewaters such as those from anaerobic digestion or landfill leachate. A wide range of media is used in these reactors, including blast furnace slag (4–6 cm), pebbles, gravel, biochar and granular activated carbon (Xia *et al.*, 2012; Forbis-Stokes *et al.*, 2018). Smaller mineral media such as sand are generally used for denitrification, although organic substrates can also be used in dentirfying systems where they can act as an electron donor as well as an effective support medium for the biofilm (Lopez-Ponnada *et al.*, 2017; Forbis-Stokes, 2018).

Biological aerated filters are well adapted for use for denitrification with a number of commercially available systems. For example, Severn and Trent have developed the Denite™ process, with the main filter made up of a 500-mm layer of support gravel on which a deep 1.8 m layer of 2–3 mm sand is placed. However, as a tertiary treatment step, methanol is normally used as an extra organic substrate. Biological Aerated Filters (BAFs) are used in a down-flow mode and can achieve removal rates of between 0.25 and 0.5 kg $NO_x$-N $m^{-2}$. Backwashing is required, consisting of an air scour followed by air-water scouring that releases trapped nitrogen gas, as well as controls solids build up within the interstices of the sand.

Anoxic RBCs are very effective at removing up to 90% of the nitrate from treated wastewaters at C/N ratios of $\leq 1.5$ with relatively short hydraulic retention times (Cortez *et al.*, 2009). In RBCs, this requires the BOD to be less than 30 mg $L^{-1}$ or 5 mg $L^{-1}$ of soluble BOD (sBOD) (Brenner *et al.*, 1984), with expected ammonium removal rates of between 1.0 and 1.5 g $m^2d^{-1}$. The degree of disc immersion is a key factor in RBC denitrification, with complete submersion being the most effective (Teixeira and Oliveira, 2001).

### 7.2.3 *Anaerobic Ammonium Oxidation*

The use of anaerobic ammonia oxidation is now found in many FFR applications, including IFAS reactors and RBCs (Wang *et al.*, 2018a; 2018b). However, the anammox process is still used as a stand-alone system for strong ammonium wastewaters such as landfill leachate (Shalini and Joseph, 2012), although it requires a constant supply of nitrite with ammonium to achieve full nitrogen removal (van Donogen *et al.*, 2001).

SHARON is an acronym for **S**ingle reactor system for **H**igh activity **A**mmonium **R**emoval **O**ver **N**itrite and is a process that produces the exact feed required for the anammox bacteria housed in a separate reactor. It is a partial nitrification process able to provide the exact mixture of ammonium and nitrite (approximately 50:50 ammonium and nitrite), with pH control used to adjust the balance (van Donogen *et al.*, 2001).

$$NH_4^+ + HCO_3^- + 0.75O_2 \rightarrow 0.5NH_4^+ + 0.5NO_2^- + CO_2 + 1.5H_2O.$$

SHARON/ANAMMOX uses two separate reactors in series, however, there are alternative processes that use a single reactor including *simultaneous partial nitrification, anammox and denitrification* (SNAD) (Chen *et al.*, 2009), *single stage nitrogen removal using ANAMMOX and partial nitritation* (SNAP) (Lieu *et al.*, 2006), *oxygen-limited autotrophic nitrification-denitrification (OLAND)* (Pynaert *et al.*, 2004), and *completely autorophic nitrogen removal over nitrite (CANON)* (Sliekers *et al.*, 2002).

Partial nitratation is achieved in oxygen limited autotrophic nitrification-denitrification RBCs (ORLAND), which are single-disc units used to treat low C:N ratio wastewaters rich in ammonium (Vlaeminck *et al.*, 2012). The process oxidizes about half of the ammonium to nitrite in the outer aerobic biofilm zones, while in the anoxic zone, the anammox bacteria reduce this and the residual ammonium predominately to nitrogen gas. However, in the process, partial nitrification must be maintained because if nitrate is produced from allowing full nitrification to occur, then overall nitrogen removal is reduced as the activity of the anammox bacteria is reduced. Close control of the nitratation rate is required, and this is based on the atmospheric exposure percentage of the discs rather than dissolved oxygen concentration (Courtens *et al.*, 2014).

Both MBBRs and IFAS systems can successfully employ partial nitratation and anammox for nitrogen removal (Fig. 7.2), and this has been reviewed by Ødegaard (2006) and di Biase *et al.* (2019).

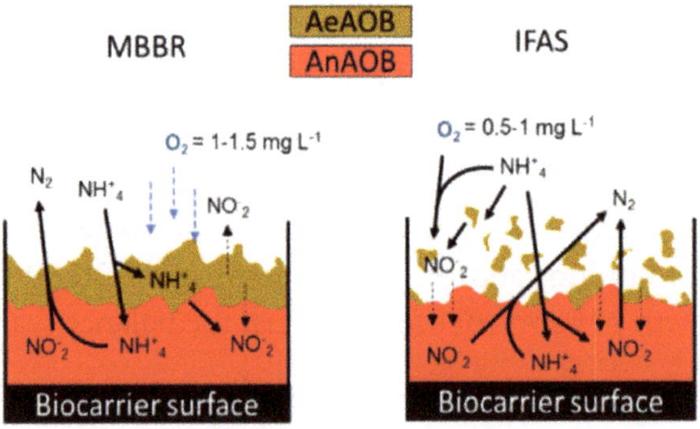


**Fig. 7.2.** Schematic comparison of the cometabolism between aerobic ammonia oxidizing bacteria (AeAOB) and anammox bacteria (AnAOB) in MBBR and IFAS systems (di Biase *et al.*, 2019). Reproduced with permission of Elsevier Publishing.

## References


Barnes, D. and Bliss, P.J. (1983) *Biological Control of Nitrogen in Wastewater Treatment*. Spon, London.

Boller, M. and Gujer, W. (1986) Nitrification in tertiary trickling filters followed by deep-bed filters. *Water Research*, **20**, (11), 1363–1373.

Brenner, R.C., Heidman, J.A., Opatken, E.J. and Petrasek, A.C. (1984) Design information on rotating biological contactors. EPA-600/2-84-106. US Environmental Protection Agency, Cincinnati, Ohio.

Burell, P.C., Keller, J. and Blackall, L.L. (1998) Microbiology of a nitrite oxidizing bioreactor. *Applied Environmental Microbiology*, **64**, (5), 1878–1583.

Cao, S., Du, R., Peng, Y., Li, B. and Wang, S. (2019) Novel two stage partial denitrification (PD)-Anammox process for tertiary nitrogen removal from low carbon/nitrogen (C/N) municipal sewage. *Chemical Engineering Journal*, **362**, 107–115.

Chen, H., Sitong, L., Yang, F., Xue, Y. and Wang, T. (2009) The development of simultaneous partial nitrification, anammox and denitrification (SNAD) process in a single reactor for nitrogen removal. *Bioresource Technology*, **100**, 1548–1554.

Cortez, S., Teixeira, P., Oliveira, R. and Mota, M. (2009) Effect of operating parameters on denitrification in an anoxic rotating biological contactor. *Environmental Technology*, **30**, (30), 1381–1389.

Courtens, E.N.P., Boon, N., De Clippeleir, H., Berckmoes, K., Mosquera, M., Seuntjens, D. and Vlaeminck, S.E. (2014) Control of nitratation in an oxygen-limited autotrophic nitrification/denitrification rotating biological contactor through disc immersion level variation. *Bioresource Technology*, **155**, 182–188.

di Biase, A., Kowalski, M.S., Devlin, T.R. and Oleszkiewicz, J.A. (2019) Moving bed biofilm reactor technology in municipal wastewater treatment: A review. *Journal of Environmental Management,* **247**, 849–866.

Eding, E.H., Kamastra, A., Verreth, J.A.J., Huisman, E.A. and Klapwijk, A. (2006) Design and operation of nitrifying trickling filters in recirculating agquiculture: A review. *Aquacultural Engineering*, **34**, (3), 234–260.

Ekama, G.A. and Wentzel, M.C. (2008). Nitrogen removal. In *Biological Wastewater Treatment: Principles, Modelling and Design* (eds. M. Henze, M.C.M. van Loosdrecht, G.A. Ekama and D. Brdjanovic). IWA Publishing, London, UK, pp. 87–138.

Forbis-Stokes, A.A., Rocha-Melogno, L. and Deshusses, M.A. (2018) Nitrifying trickling filters and denitrifying bioreactors for nitrogen management of high strength anaerobic digestion effluent. *Chemosphere*, **204**, 119–129.

Fuerst, J.A. and Sagulenko, E. (2011) Beyond the bacterium: Planctomycetes challenge our concepts of microbial structure and function. *Nature Reviews Microbiology*, **9**, 403–413.

Godoy-Olmos, S., Martínez-Llorens, S., Tomás-Vidal, A. and Jover-Cerdá, M. (2016) Influence of filter medium type, temperature and ammonia production on nitrifying trickling filters performance. *Journal of Environmental Chemical Engineering*, **4**, 328–340.

Gutierrez-Wing, M.T. and Malone, R.F. (2006) Biological filters in aquaculture: Trends and research directions for freshwater and marine applications. *Agricultural Engineering*, **34**, (3), 163–171.

Kartal, B., Maalcke, W.J., de Almeida, N.M., Cirpus, I., Gloerich, J., Geerts, W., den Camp, H.J.M.O., Harhangi, H.R., JanssenMegens, E.M., Francoijs, K.-J., Stunnenberg, H.G., Keltjens J.T., Jetten, M.S.M. and Strousm, M. (2011) Molecular mechanism of anaerobic ammonium oxidation. *Nature*, **479**, 127–130

Lackner, S., Gilbert, E.M., Vlaeminck, S.E., Joss, A., Horn, H. and van Loosdrecht, M.C.M. (2014). Full-scale partial nitritation/anammox experiences — an application survey. *Water Research*, **55**, 292–303.

Lekang, O.-I. and Kleppe, H. (2000) Efficiency of nitrification in trickling filters using different filter media. *Aquacultural Engineering*, **21**, 181–199.

Li, N., Zeng, W., Yang, Y., Wang, B., Li, Z. and Peng, Y. (2019) Oxygen mass transfer and post-denitrification in a modified rotating drum biological contactor. *Biochemical Engineering Journal*, **144**, 48–56.

Lieu, P.K., Homan, H., Kurogi, A., Kawagoshi, Y. and Furukawa, K. (2006) Characterization of sludge from single stage nitrogen removal using anammox and partial nitritation (SNAP). *Japanese Journal of Water Treatment Biology*, **42**, 53–64.

Lopez-Ponnada, E.V., Lynn, T.J., Peterson, M., Ergas, S.J. and Mihelcic, J.R. (2017) Application of denitrifying wood chip bioreactors for management of residential non-point sources of nitrogen. *Journal of Biological Engineering*, **11**, (16). https://doi.org/10.1186/s13036-017-0057-4

Malone, R.F. and Pfeiffer, T.J. (2006) Rating fixed film nitrying biofilters used in recirculating aquaculture systems. *Agricultural Engineering*, **34**, (3), 389–402.

Ødegaard, H. (2006) Innovations in wastewater treatment: The moving bed biofilm reactor process. *Water Science and Technology*, **53**, (9), 17–33. https://doi.org/10.2166/wst.2006.284

Persson, F., Wik, T., Sörensson, F. and Hermansson, M. (2002) Distribution and activity of ammonia oxidizing bacteria in a large full-scale trickling filter. *Water Research*, **36**, 1439–1448.

Pynaert, K., Smets, B.F., Beheydt, D. and Verstraete, W. (2004) Start-up of autotrophic nitrogen removal reactors vis sequential biocatalyst addition. *Environmental Science and Technology*, **38**, 1228–1235.

Shalini, S.S. and Joseph, K. (2012) Nitrogen management in landfill leachate: Application of SHARON, ANAMMOX and combined SHARON-ANAMMOX process. *Waste Management*, **32**, (12), 2285–2400.

Sliekers, A.O., Derwort, N., Gomez, J.L., Strous, M. Kuenen, J.G. and Jetten, M.S. (2002) Completely autotrophic nitrogen removal over nitrite in one single reactor. *Water Research*, **36**, 2475–2482.

Teixeira, P. and Oliveira, R. (2001) Denitrification in a closed rotating biological contactor: Effect of disk submergence. *Process Biochemistry*, **37**, 345–349.

van Donogen, U., Jetten, M.S. and van Loosdrecht, M.C. (2001) The SHARON-Anammox process for treatment of ammonium rich wastewater. *Water Science and Technology*, **44**, (1), 153–160.

Vlaeminck, S.E., De Clippeleir, H. and Verstraete, W. (2012) Microbial resource management of one-stage partial nitritation/anammox. *Microbial Biotechnology*, **5**, (3), 433–448.

Wagner, M. and Amman, R.I. (1998) Identification and activities *in situ* of *Nitrosospira* and *Nitrospira* spp. as dominant populations in a nitrifying fluidized bed reactor. *Applied Environmental Microbiology*, **64**, (9), 3480–3485.

Wang, C., Liu, S., Xu, X., Zhang, C., Wang, D. and Yang, F. (2018a) Achieving mainstream nitrogen removal through simultaneous partial nitrification, anammox and denitrification process in an integrated fixed film activated sludge reactor. *Chemosphere*, **203**, 457–466. https://doi.org/10.1016/j.chemosphere.2018.04.016

Wang, D., Wang, G., Yang, F., Liu, C., Kong, L. and Liu, Y. (2018b) Treatment of municipal sewage with low carbon-to-nitrogen ratio via simultaneous partial nitrification, anaerobic ammonia oxidation, and denitrification (SNAD) in a non-woven rotating biological contactor, *Chemosphere*, **208**, 854–861. https://doi.org/10.1016/j. chemosphere. 2018.06.061

Wik, T. (1999) Adsorption and denitrification in nitrifying trickling filters. *Water Research*, **33**, (6), 1500–1508.

Xia, M., Tao, W., Wang, Z. and Pei, Y. (2012) Treatment of anaerobically digested dairy manure in a two-stage biofiltration system. *Water Science and Technology*, **65**, (11), 1975–1981. https://doi.org/10.2166/wst.2012.097

# 8
# Biological Air Treatment

## 8.1 Introduction

There are many methods of removing volatile and odorous compounds from contaminated air and gaseous emissions. These include thermal oxidation, adsorption using activated carbon, ozonation, chemical treatment, or if the contaminant is non-toxic then masking (Barbusinski *et al.*, 2017). However, biological treatment offers a reli able and simple-to-operate system with low capital and operating costs without any significant waste product. Although waste gases can be pumped into an activated sludge reactor to remove low levels of contaminant, specially constructed biological air treatment units are generally required.

Although the principle of biological air treatment is similar for all reactor systems, three different types of processes — biofilters, bioscrubbing and biotrickling filters — are available, though there is often much confusion in the precise use of these terms. There are many different designs within each category, but all operate on the same principle of the gaseous compound(s) being absorbed into the water by gas scrubbing or humidification, followed by a biological treatment phase where bacteria degrade the contaminants. The clean, decontaminated air is released into the atmosphere, while contaminants are biodegraded. In summary, the three main biological air treatment processes are:

*Biofiltration:* The contaminated air is moistened after going through a humidifier or scrubber, specifically for that purpose. The humid air is then passed upward through a filter comprising natural medium (e.g., woodchip) on which the biofilm develops that degrades the contaminants (Sec. 8.2).

*Bioscrubbing:* This is the classic configuration similar to normal gas scrubbers, where the contaminated air is forced upward through the inert medium while wash water containing bacterial flocs is sprayed onto the surface of the medium from the top. This counter-current action of air and water ensures that the contaminants become fully absorbed into the liquid phase. The scrubber is primarily used to remove the contaminants, and any treatment that takes place is incidental. The wash water passes through the scrubber into an activated sludge unit, a moving bed biofilm reactor (MBBR), or an integrated fixed-film activated sludge (IFAS) system, where any remaining contaminant is biodegraded. The wash water is taken from the secondary settlement tank that seeds it with the bacteria and nutrients needed for treatment and to support the scrubber action. This system is preferred where gas emissions requiring treatment are more irregular and where high volumes of low level contaminated air need treatment (Sec. 8.3).

*Biotrickling filtration:* These are single treatment units with gaseous absorption into the liquid phase and biological treatment taking place simultaneously. This is a closed system with the wash water constantly recirculated, although a nutrient feed is normally required. The media is usually ceramic (or plastic) and inert, so it does not require replacement every few years, which is the case with other designs using organic-based media (Sec. 8.4).

The three processes are compared in Table 8.1.

Fluidized beds are also used for biological air treatment with the advantage that the biofilm developing on the very small media produces a homogenous community structure than the other types of filters used. Apart from traditional inert media, both peat and sawdust have been successfully used to treat volatile organic carbons (VOCs) (Clarke *et al.*, 2008; Delebarre *et al.*, 2007), although the media should be pre-activated with microorganisms using activated sludge.

**Table 8.1.** Comparison of advantages and disadvantages of the fixed-film systems' biofiltration and biotrickling filters with bioscrubbers (Malakar *et al.*, 2017). Reproduced with permission of Elsevier Publishing.

| Category | Biofilter | Biotrickling filter | Bioscrubber |
|---|---|---|---|
| Characteristics/ Application | ➢ Immobilized biomass<br>➢ Immobile water phase<br>➢ Single reactor<br>➢ Removal of odor and low VOC concentrations<br>➢ Target VOC concentration is less than 1 g m$^{-3}$ | ➢ Immobilized biomass<br>➢ Mobile water phase<br>➢ Single reactor<br>➢ Low / medium VOC concentrations<br>➢ Target VOC concentration is less than 0.5 g m$^{-3}$ | ➢ Suspended biomass<br>➢ Mobile water phase<br>➢ Two reactors<br>➢ Low / medium VOC concentrations<br>➢ Target VOC concentration less than 5 g m$^{-3}$ |
| Advantages | ➢ High gas-phase surface area<br>➢ Easy operation and start-up<br>➢ Low capital and operation costs<br>➢ Use of natural media in biofilter bed<br>➢ Does not need inoculation<br>➢ Nutrients need not be provided<br>➢ Does not produce waste<br>➢ Degrades a wide range of components<br>➢ Easy to operate and maintain<br>➢ No unnecessary waste streams are produced<br>➢ Low pressure drop | ➢ Better retention to slow growing microorganisms<br>➢ High efficiency<br>➢ Does not produce wastewater<br>➢ High durability of synthetic media<br>➢ Less operating and capital constraints<br>➢ Short retention time / high volume through put<br>➢ Capability to treat acid degradation product of VOCs | ➢ Better control of reaction conditions<br>➢ Compact equipment<br>➢ Low pressure drop<br>➢ Hot polluted air is treatable<br>➢ Does not tolerate clogging<br>➢ Does not tolerate drying<br>➢ Able to deal with high flow rates and severe fluctuations<br>➢ Operational stability and better control of operating parameters<br>➢ Relatively low pressure drop<br>➢ Relatively smaller space requirements |

(*Continued*)

**Table 8.1.** (*Continued*)

| Category | Biofilter | Biotrickling filter | Bioscrubber |
|---|---|---|---|
| Disadvantages | ➢ Poor control of operation conditions<br>➢ Slow adaptation to variation in gas concentration<br>➢ Large area required<br>➢ Short durability of natural media in comparison to synthetic media<br>➢ BFs face problems such as clogging and drying<br>➢ Compaction of natural bed and channeling<br>➢ High-pressure drops<br>➢ Problems in treating acidic gases<br>➢ Less treatment efficiency at high concentrations of pollutants<br>➢ Packing has a limited life<br>➢ Clogging of the medium due to particulate medium | ➢ Low surface area for mass transfer<br>➢ Difficult start-up procedures<br>➢ High operation cost<br>➢ Complex operation<br>➢ Needs preparation of water and nutrients<br>➢ Synthetic supporting materials are expensive<br>➢ Accumulation of excess biomass in the filter bed<br>➢ Requirement of design for fluctuating concentration<br>➢ Complexity in construct and operation<br>➢ Secondary waste-stream | ➢ Low surface area for mass transfer<br>➢ Wash out of low growing microorganisms<br>➢ Disposal of excess sludge<br>➢ Difficult start-up procedures<br>➢ Extra air supply needed at high degradation rates<br>➢ High investments, maintenance and operational cost<br>➢ Produces wastewater<br>➢ Expensive<br>➢ Complex operation<br>➢ Treats only water soluble compounds<br>➢ Can be complicated to operate and maintain<br>➢ Extra air supply may be needed<br>➢ Excess sludge will require disposal<br>➢ Generation of liquid waste |

## 8.2 Biofiltration

Biofiltration is used primarily for the removal of common problem odors such as ammonia and hydrogen sulfide that arise from wastewater treatment, fish processing and landfill facilities. It can also be used at industrial sites to remove VOCs such as toluene, hexane, ethanol, etc. (Delhoménie and Heitz, 2005). The contaminated air is humidified before being passed upward through a biofilter containing organic media such as peat, bark, biochar, wood chip, leaf compost, or composted organic waste derived from refuse or agriculture (Delhoménie and Heitz, 2005; Jun and Xu, 2009; Dorado *et al.*, 2010; Baltrenaite *et al.*, 2017; Zhong *et al.*, 2020) (Fig. 8.1). Sand, soil or other very fine material is no longer used for biological air filtration due to becoming easily clogged. To maintain good porosity a bulking agent is often mixed with compost and non-fibrous peat such as wood chips. The media must be carefully packed to ensure even

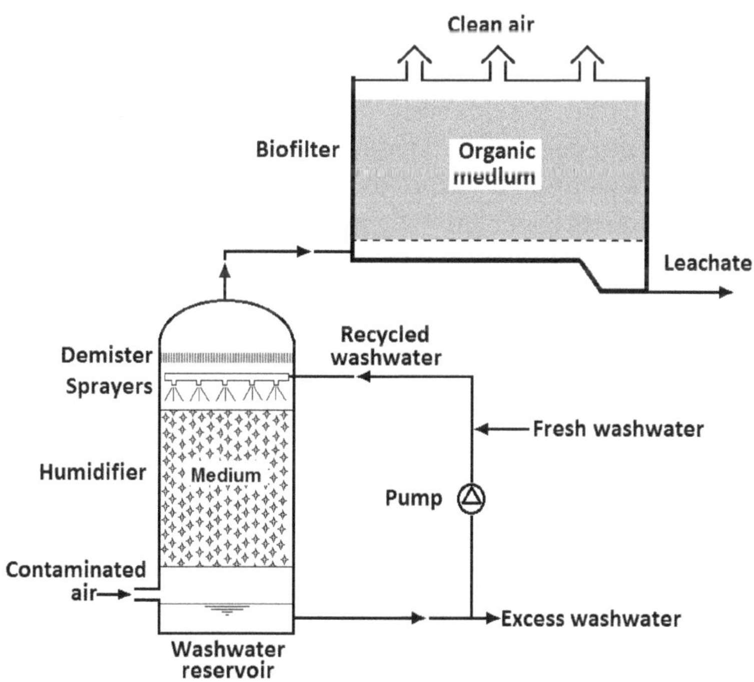


**Fig. 8.1.** Schematic of a biofilter unit with a humidifier.

consistency, thus ensuring no short-circuiting of the contaminated air occurs. If the treated air stream contains dust, this will become trapped within the interstices of the media and lead to blockages. The media generally has a low moisture content (30–40%), and so the incoming air must be as saturated with as much moisture as possible (90–95%). Like other media, the biofilm develops on the surface of the organic material, which it can metabolize for extra nutrients, although some nutrients or mineral supplements may be required over time as the media ages. Layered media can also be used in biofilters to help develop different consortia of bacteria, with readily degradable fractions of the contaminated air removed in the lower layer to allow the more recalcitrant fractions to be removed in the following layers where more specialized, often slower growing, microorganisms can thrive.

The biofilters can be operated as a single unit or in parallel depending on the volume of gas, the concentration of the contaminants, and their breakdown rates. Modern biofilters are generally enclosed for protection from the weather and to aid temperature control and retention time (known as the EBRT, or empty bed residence time). Earlier systems were usually open to the atmosphere, rather like normal percolating filters, with the contaminated air introduced at the base through a series of perforated pipes. The pH of the biofilter should remain within the range of 7.0–8.0 to ensure optimal microbial activity and temperate in the mesophilic range of 20–30°C (Williams and Miller, 1992a; 1992b). Where the contaminated airflow is acidic, dolomite can be added to the media to help reduce acidification, or chemical scrubbing may be required. Excessive ammonia in the air stream is the most common cause for acidification within biofilters, and loadings exceeding 10 mg N-$NH_3$ $m^{-3}$ should be avoided. The media offers the extra adsorptive capacity to help retain the contaminants, which allows the microorganisms to then biodegrade them. Normally, the compost-based media has sufficient biodiversity of microorganisms, but bark or wood chip will need to be seeded with activated sludge or a proprietary microbial supplement. Loading rates from the humidifier to the filter are between 50 and 300 $m^3m^{-3}$ $h^{-1}$ at a reactor height of 1.5 m, thus ensuring a residence time of 30–60 seconds. The organic media has a limited life and should be renewed every 1–5 years depending on the contaminants and the media. The life span of organic media in biofilters is significantly reduced by many VOCs during treatment. Lebrero *et al.* (2014), in a comparative study of three different filter media, found that exposure to toluene for 21 days increased their biodegradability by 26% in bark, 20% in compost, and 17% in macadamia nutshells. They suggest that potential media be tested before selection, for the effect of the VOCs to be treated on the structure of the media.

While the pressure within the reactor should be at normal atmospheric pressure, any increase in the energy required to pump the contaminated air through the filter may indicate compaction of the media as it breaks down, or any reduction may be due to air channeling. Both scenarios indicate that the media should be replaced. Used media is generally used for landfill cover and is not suitable for reuse in agriculture or horticulture.

Biofilters can treat hazardous odors (e.g., $H_2S$ and $NH_3$), VOCs and greenhouse gas emissions (e.g., $CH_4$), and a wide range of microorganisms are involved in their biodegradation (Table 8.2). Diffuse methane emissions from landfill, anaerobic digestion units and piggeries have been a cause of concern due to its high global warming

**Table 8.2.** Example of the range of microorganisms found as part of the biofilm community that can biodegrade odorous sulfur compounds (Kennes and Veiga, 2004). Reproduced with permission of Elsevier Publishing.

| Compounds | Microorganisms | Other characteristic |
|---|---|---|
| Hydrogen sulphide | *Thiobacillus* sp. CHI11 | Autotrophic |
| | *Thiobacillus thioparus* TK-m | Autotrophic |
| | *Thiobacillus thiooxidans* KS1 | Autotrophic |
| | *Hyphomicrobium* sp.I55 | Methylotrophic |
| | *Xanthomonas* sp. DY44 | Heterotrophic |
| | Unidentified basidiomycete | Fungus |
| | *Cephalosporium* sp. | Fungus |
| Dimethyl disulphide | *Thiobacillus* sp. | Autotrophic |
| | *Hyphomicrobium* sp.I55 | Methylotrophic |
| | *Pseudomonas acidovorans* DMRll | Heterotrophic |
| | Unidentified basidiomycete | Fungus |
| | *Cephalosporium* sp. | Fungus |

potential. The problem is that over half of all methane emissions are diffuse and have a concentration of <3% (v/v). This makes treatment by conventional physical-chemical scrubbing difficult and ineffective. Biofiltration is a cheap and highly effective solution that can be used *in situ* where the methane is being generated and normally lost to the atmosphere. Biofilters have been used to capture and biodegrade the gas in many different scenarios (La *et al.*, 2018), but problems have been encountered due to its poor mass transfer from the contaminated air to the biofilm. Methanotrophic activity is temperature dependent, as is methane and oxygen diffusion coefficients, with optimum performance being between 25 to 35°C (Fedrizzi *et al.*, 2018). Expanded vermiculite is an ideal media for methanotrophs in the treatment of methane (Brandt *et al.*, 2016). The effect of temperature, pH, media and loading rates are reviewed in terms of operational management by La *et al.* (2018), as is the biology of the process.

Fungal biofilters are more effective than bacteria-only systems in dealing with methane and other hydrophobic contaminants. In part, this is due to the fungal hyphae growing into the interstices of the medium and so into the gas phase, significantly increasing the potential surface contact area with the contaminant. Fungal hyphae also have cell surface-associated hydrophobins, which are proteins that confer a hydrophobic property to the cell surface, helping the uptake of methane onto aerial hyphae, achieving 90% removal efficiency (Lebrero *et al.*, 2016). Fungi play an important role in the biofilter community structure, especially in terms of resilience due to their ability to withstand desiccation and acidification, which are common stressors in biofiltration. Fungal contribution to total biomass is found to be greatest in the most desiccation-prone areas of biofilters, thus making them an important member of the microbial community of all biofilters (Oliver *et al.*, 2016). Their hyphae, which create extensive networks between the organic media particles, are rich in adsorption sites improving the scavenging of contaminants as the air passes through the filter. Fungi also have special oxidative capacities that support cometabolism, with aromatic pollutants and lignin facilitating the uptake of hydrophobic VOCs (van Groenestijn *et al.*, 2001; Kennes and Veiga, 2004) (Table 8.3).

**Table 8.3.** Examples of fungi able to degrade and grow on some volatile aliphatics and phenolic pollutants (Kennes and Veiga, 2004). Reproduced with permission of Elsevier Publishing.

| Compounds | Organisms |
|---|---|
| *n*-Alkanes | |
| $C_2$-$C_4$-Alkanes | *Acremonium* sp. |
| $C_6$-$C_{19}$-Alkanes | *Cladosporium resinae* ATCC 22711 |
| *n*-Butane | *Graphium* sp. ATCC 58400 |
| $C_1$-$C_9$-Alkanes | *Scedosporium* sp. |
| *n*-Hexadecane | *Trichosporon, veenhuisii* CBS 7136 |
| Phenols | |
| Phenol | *Aspergillus japonicus* |
| Phenol | *Penicillium* spp. |
| Phenol | *Trichosporon cutaneum* |
| Phenol | *Trichosporon guehoae* CBS |
| *m*-Cresol | 8521 |
| Phenol | *Trichosporon·veenhuisii* |
| *m*-Cresol | CBS 7136 |
| *p*-Cresol | *Aspergillus fumigatus* ATCC 28282 |
| *o*-Cresol | *Penicillium.frequentans* ATCC 96048[a] (growth on glucose and phenol) |
| *o*. *m*, and *p*-Cresol | *Trichosporon cutaneum* KUY-6A |
| Others | |
| Formaldehyde Methanol | *Gliocladium deliquesecens* |
| Formaldehyde Methanol | *Paecilomyces mriotii* |
| Diethyl ether | *Graphium* sp. ATCC 58400 |
| Methyl *tert*-butyl ether | *Graphiun* sp. ATCC 58400[a] (growth on *n*-alkanes) |

[a]Cometabolism.

Careful media selection is important to achieve efficient ammonia removal by biofiltration. Compost and sludge have been widely used as media achieving ammonia removal efficiencies of 97–99% and 95–99%, respectively (Jun and Xu, 2009). These materials already contain a diverse microbial community and have high water retention capabilities and high nutrient content. Composting and biofiltration are often used together, with the composted material being used as the media for the biofilter (Zhong *et al.*, 2020).

Hydrogen sulfide ($H_2S$) is a particularly hazardous and odorous compound with respiratory problems encountered at concentrations of 2 ppm and above. The main danger is that as the concentration of gas increases, it paralyzes the sense of smell, thus making people unaware of its presence. Included in the list of suspected disorders from exposure to $H_2S$ are liver cirrhosis, cancer, Alzheimer's disease, diabetes and Down's syndrome. Biofiltration is an efficient treatment system for the capture and biodegradation of $H_2S$, providing an ideal environment for microbial degradation by bacteria such as *Thiobacillus ferrooxidans*, *Thiobacillus thiooxidans*, *Thiobacillus novellus*, *Thiobacillus thioparus*, *Thiobacillus denitrifican*, *Thermothrix azorensis*, *Thiothrix nivea*, *Thioalkalispira microaerophila* and *Thiomicrospora frisia*. The media selected must have a large buffering capacity, and be mechanically resistant, and include inert media such as expanded schist, acid-resistant polyethylene, concrete waste and organic media such as wood chips, bark and shells. The microbiology of the process and the effects of operational parameters on performance are reviewed by Vikrant *et al.* (2018).

Biofiltration is very effective for odor control and removal of VOCs, achieving between 70–95% efficiency rates. There will be a small amount of wastewater discharged from the filter that requires further treatment either on-site or via a sewer connection. With low capital and operational costs, their simplicity of operation and low maintenance makes them a popular choice for biological air treatment.

## 8.3 Bioscrubbing

Bioscrubbers were developed in Germany in the early 1970s and are often confused with biotrickling filters. However, they are completely different processes and should be thought of as suspended-growth bioscrubbers and fixed-growth bioscrubbers, respectively. So, unlike biofiltration and biotrickling filters, which are biological fixed-film systems, bioscrubbers are a two-unit operational process with the physical absorption of VOCs into wash water in an enclosed media packed tower (i.e., the scrubber), followed by a dedicated biological treatment process of usually activated sludge, although MBBRs and IFASs can be used. The contaminated air is pumped up through the medium packed tower with a counter current of clean wash water containing bacteria and some nutrients that strip the contaminants from the gas phase into solution. This results in a constant discharge of contaminated water, which requires further treatment biologically (van Groenestijn, 2001) (Fig. 8.2). The thin biofilm on the inert media and the bacteria in the water phase contribute to some biodegradation of the contaminants, now in solution, but this is minor compared to the biological treatment step. The fresh wash water comes from the final effluent and is rich in bacteria and nutrients, which helps supports a thin biofilm within the scrubber.

Bioscrubbing requires the contaminant to be both water soluble and degradable by the activated sludge microbial community. These units can capture and treat a wide range of suitable compounds,

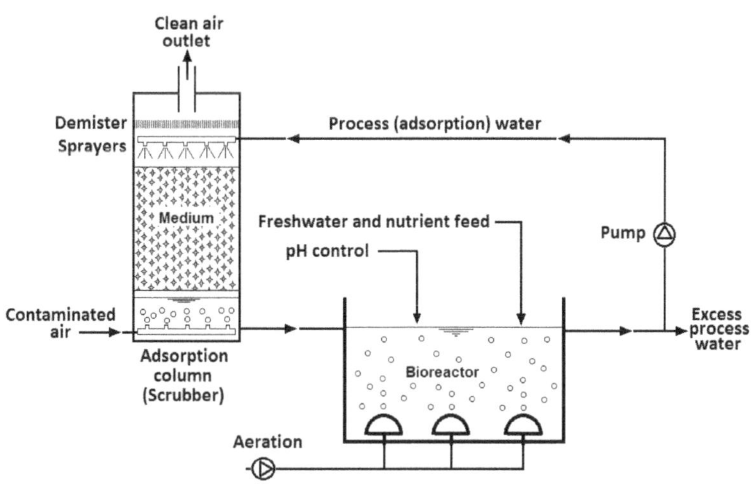


**Fig. 8.2.** Schematic of a simple bioscrubber used in conjunction with activated sludge.

including alcohols, aldehydes, sulfur and nitrogen compounds, mercaptans, and many other toxic and odorous compounds, including ammonia and hydrogen sulfide. Expected removal efficiencies are 80–95% for ammonia, 80–90% for VOCs, and 70–80% for odors. For non-toxic odors the removal efficiency required is closely linked to the compound's odor threshold. Bioscrubbers are ideal, where pH control is required as alkaline or acidic adjustment can be made to the wash water, but the process is best suited to treating highly soluble compounds (e.g., aldehydes, alcohols, fatty acids).

Bioscrubbers provide better pH control than either biofilters or biotrickling filters, are more compact, do not have media problems such as ponding or media degradation, provide a cleaner final effluent, and are easier and more reliable to operate. The biological phase will need nutrient supplements if it is a dedicated unit for the bioscrubber, although if connected to the main treatment unit (e.g., a bioscrubber treating contaminated air at a wastewater treatment plant) then sufficient nutrients will be available. On startup, the bioreactor will need seeding with microorganisms. If activated sludge is used, then it will take 3–6 weeks for the bacteria to acclimatize to their new carbon substrate, depending on its biodegradability. Ideally, an inoculum should be obtained from another bioscrubber treating the same contaminants or a commercial seed purchased. There is a possibility that during this start-up phase, incomplete biodegradation may lead to contaminants being stripped from the bioreactor by the aeration device. This can also occur due to overloading or operational factors such as a fall in temperature leading to reduced removal rates. In these situations, the air from the bioreactor should be recycled through the bioscrubber, making it a closed loop system.

Biofiltration and biotrickling filters are restricted to industries where there is a constant emission of the contaminant(s). They are probably unviable for situations where short work shifts or irregular work patterns exist. Ideally, they require a work schedule resulting in emissions for a minimum of 8–10 hours per day, at least 5 days a week. However, bioscrubbers are more flexible in terms of intermittent loading, although the bioreactor will need a sufficient input of an alternative organic substrate to maintain active microbial biomass when their normal carbon source (i.e., VOCs) is not being captured.

Bioscrubbers are usually stainless steel columns that are 1.5–2.0 m in diameter and 10–12 m in height. They are operated at very short gas retention times of 1 second, with higher resultant gas loadings (3,000–4,000 $m^3m^{-2}h^{-1}$) than biofilters (100–150 $m^3m^{-2}h^{-1}$), making the process much more compact and cheaper to construct. Where they use a dedicated biological treatment unit, this tends to be located adjacent to the column and is quite small ($<25\ m^3$) in volume. Hansen and Rindel (2000) give an example of a bioscrubber treating the odorous air from the preliminary treatment area at the Damhusaaen Waste Water Treatment Plant (WWTP) in Copenhagen, where hydrogen sulfide and other sulfur compounds are treated using sodium hydroxide to neutralize the sulfuric acid formed.

## 8.4 Biotrickling Filtration

Biotrickling filters are true fixed-film reactors and combine gas scrubbing and biological treatment into a single unit. The enclosed filter is filled with an inert media on which a biofilm develops that decomposes the contaminants in the gas phase. Wash water is sprayed evenly over the surface of the media, ensuring that the biofilm is kept wet and that the interstices are moist to ensure the maximum absorption of contaminants from the counter-flow of contaminated air, which is pumped into the reactor from the base and operated at atmospheric pressure (Fig. 8.3). The VOCs and other contaminants are the carbon source for the bacteria, but as this is a closed loop system, nutrient supplements are required to ensure an active and healthy biofilm. Although the biofilm is relatively thin, it can become problematic and can cause localized ponding, channeling within the filter, and a loss of treatment efficiency. Therefore it is important to ensure that the filter is not organically overloaded. Excess biomass is constantly detached from the media, creating excess solids that can also eventually block the filter and so must be periodically removed. Conductivity, solids and pH are closely monitored and corrected when necessary, and the temperature of the flue gas is also monitored to maintain the optimum temperature for the

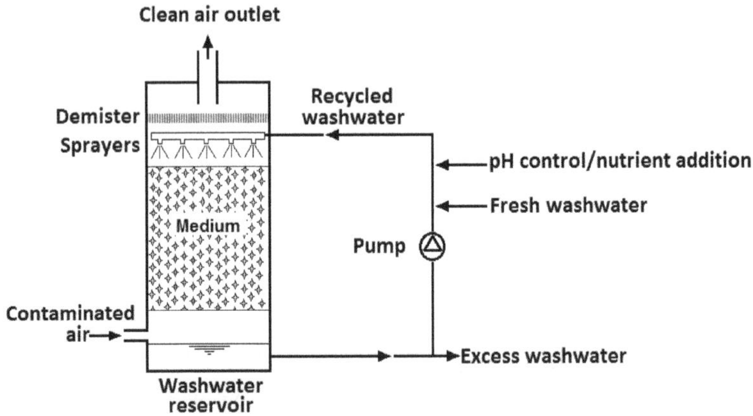


**Fig. 8.3.** Schematic of a biotrickling filter.

biofilm within the reactor (30–35°C). The residence time or the EBRT is very short, just tenths or even hundreds of a second, although this will decrease if biofilm accumulation increases (Rybarczyk *et al.*, 2019). Biotrickling filters need a constant supply of carbon for the biofilm, and so are not suitable for short working periods or intermittent use. The process does produce some excess sludge, and any excess wash water must be disposed into the sewer.

As with biofilters, biotrickling filters have to be inoculated with a suitable consortium of microorganisms upon startup. Unlike percolating filters, they are unable to develop their own microbial flora without help. The best scenario is to use activated sludge to inoculate a new system that provides the widest selection of potential species, primarily bacteria and fungi. As with biofilters, fungi can be important in the treatment of VOCs, especially those that are hydrophobic (Sec. 8.2). Adaptations and growth will take several weeks after initial inoculation with activated sludge, although this can be speeded up by using commercial bioaugmentation or by obtaining inocula from biotrickling filters treating the same contaminants (Zhou *et al.*, 2016). Rybarczyk *et al.* (2019) reviewed the microbial ecology and dominant microbial species found in biotrickling filters treating a wide variety of VOCs.

Inert media is preferred, and a wide variety of plastic and ceramic media has been used either singly or in layers, including polyurethane foam cubes, perlite, lava rock, plastic pall rings, ceramic raschig rings and clay balls. Some organic materials are occasionally used, and these include walnut shells, which are very resistant to biodegradation, and mussel shells, which can be dissolved in acidic environments. High porosity (35–40%) is required in the media to ensure good distribution of the contaminated air, along with a high specific surface area for maximum biofilm growth to maximize the mass transfer of contaminants to the microbial layer, sufficient-sized interstices and voidage to prevent pressure drop and allow good drainage and air movement, and to promote high water retention to prevent drying. A number of different configurations of biotrickling filter are available, most of which are counter-current flow, with one or more towers used in series (Rybarczyk *et al.*, 2019) (Fig. 8.4). Where high biofilm growths are problematic, then co-current designs have been considered but are not the norm. A compact co-current biotrickling filter has been successfully developed and used for odor control at small-scale livestock farms by Kawase *et al.* (2014). In pilot-scale experiments using biotrickling filters for anoxic desulphurization, Almenglo *et al.* (2016) compared counter- and co-current gas-liquid operations and found that the counter-current operation produced better performance. However, they found that switching from one mode to the other redistributed biomass ensured good operating conditions.

Biotrickling filters have been used to treat a wide range of VOCs and odors, especially acidic emissions, including fatty acids and their esters, aldehydes, ketones and alcohols, naphthalene, styrene, phenols, nitrogen compounds (e.g., $NH_3$, amines), sulfur components (e.g., $H_2S$, mercaptans, carbon disulphide), and chlorine components (e.g., trichloroethene, 1,2-dichloroethane, chlorophenol, dichloromethane, monovinylchloride). They are well suited to all water soluble VOCs and other gaseous contaminants, and can treat high concentrations of VOCs (400–2,000 mg $m^3$), mercaptans (5–100 mg $m^3$), $H_2S$ (50–200 mg $m^3$), $NH_3$ (100–400 mg $m^3$), and odors (up to 25,000 OU $m^3$), with removal efficiencies of 80–95% for ammonia, hydrogen sulfide and hydrocarbons, and 70–90% for mercaptans at standard temperature and pressure (25°C and 1 kPa). Odor reductions are up to 90% and are usually below detection levels, as expressed by the odor number.

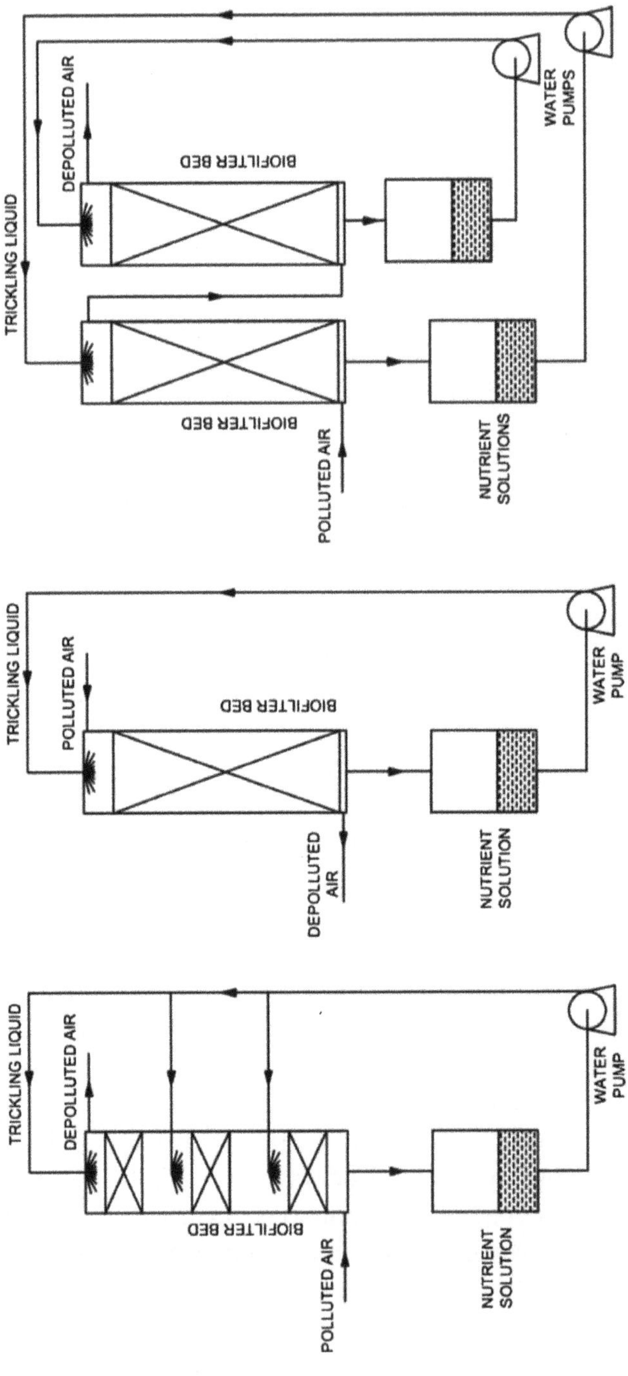


**Fig. 8.4.** There are many different configurations of biotrickling filter configurations. For example: (a) Counter current three-layer biotrickling filter, (b) co-current biotrickling filter, and (c) two-stage biotrickling filter (Rybarczyk *et al.*, 2019). Reproduced with permission of Elsevier Publishing.

During anaerobic digestion, hydrogen sulfide ($H_2S$) can be formed from the sulfur compounds in the waste. This can result in between 0.1 and 5.0% of the gas ending up in the methane, which must be removed before the biogas can be used. Biotrickling filters have been successfully used for biogas desulphurization, with both anoxic (using $NO_2^-$ and $NO_3^-$ as electron acceptors) and aerobic biofilters (using $O_2$ as the electron acceptor) capable of removing the $H_2S$ (Vikrant *et al.*, 2018). At low loadings, $SO_4^2 - S$ is the dominant end product with the proportion of $S^o$–S increasing at higher loading rates. The microbial ecology of the process is discussed by Wu *et al.* (2020), who found that the dominant genera of sulfide oxidizing bacteria were *Pseudomonas*, *Halothiobacillus* and *Sulfurimonas*, with their dominance increasing with loading.

The oxidizing reactions are:

$$H_2S \leftrightarrow H^+ + HS^-,$$
$$HS^- + 0.5O_2 \rightarrow S + OH^-,$$
$$HS^- + 2O_2 \rightarrow SO_4^{2-} + H^+.$$

There is growing interest in using highly efficient species to optimize the biotrickling filter desulphurization performance. Juntranapaporn *et al.* (2019) immobilized the chemolithoautotrophic bacterium *Paracoccus pantotrophus* onto plastic media used in a pilot biotrickling filter and recorded removal rates of up to 83 g $H_2S$ $m^{-3}h^{-1}$ at a loading rate of 84 g $H_2S$ $m^{-3}h^{-1}$, a removal rate of 99.5%.

Biotrickling filters are considered superior to biofiltration systems due to smaller space requirements, increased process stability, low flow resistance within the filter, effective temperature and pH control (Schiavon *et al.*, 2016). They are reviewed by Rybarczyk *et al.* (2019).

## References

Almengio, F., Bezerra, T., Lafuente, F.J., Gabriel, D., Ramirez, M. and Cantero, D. (2016) Effect of gas-liquid flow pattern and microbial diversity of a pilot-scale biotrickling filter for anoxic biogas desulfurization. *Chemosphere*, **157**, 215–223.

Baltrenaite, E., Baltrenas, P., Bhatnagar, A., Vulppo, T., Selenius, M., Koistinen, A., Dahl, M. and Penttinen, O.-P. (2017) A multicomponent approach to using waste derived biochar in biofiltration: A case study based on dissimilar types of waste. *International Biodeterioration and Biodegradation*, **119**, 565–576.

Barbusinski, K., Kalemba, K., Kasperczyk, D., Urbaniec, K. and Kozik, V. (2017) Biological methods for odor treatment — a review. *Journal of Clean Production*, **152**, 223–241. https://doi.org/10.1016/J.JCLEPRO.2017.03.093

Brandt, E.M., Duarte, F.V., Vieira, J.P.R., Melo, V.M., Souza, C.L., Araújo, J.C. and Chernicharo, C.A.L. (2016) The use of novel packing for improving methane oxidation in biofilters. *Journal of Environmental Management*, **182**, 412–420.

Clarke, K., Hill, G.A. and Pugsley, T. (2008) Improved VOC bioremediation using a fluidized bed peat bioreactor. *Process Safety and Environmental Protection*, **86**, 283–290. https://doi. org/10.1016/J.PSEP.2008.04.002.

Delebarre, A., Andres, Y., Pellerano, M., Pero, P. and Garcia Munzer D.G. (2007) Biofiltration of volatile organic compounds by a fluidized bed of sawdust. *International Journal of Chemical Reactor Engineering*, **5**, (1). https://doi.org/10.2202/1542-6580.1402.

Delhoménie, M.-C. and Heitz, M. (2005) Biofiltration of air. *Critical Reviews in Biotechnology*, **25**, (1–2), 53–72.

Dorado, A.D., Lafuente, F.J., Gabriel, D. and Gamisans, X. (2010) A comparative study based on phsycal characteristic of suitable packing materials in biofiltration. *Environmental Technology*, **31**, (2), 193–204.

Fedrizzi, F., Cabana, H., Ndanga, E. and Cabral, A.R. (2018) Biofiltration of methane from cow barns: Effect of climatic conditions and packing bed media acclimatization. *Waste Management*, **78**, 669–676.

Hansen, N.G. and Rindel, K. (2000) Bioscrubbing, an effective and economic solution to odour control at wastewater treatment plants. *Water Science and Technology*, **41**, (6), 155–164.

Jun, Y. and Xu, W. (2009) Ammonia biofiltration and community analysis of ammonia-oxidizing bacteria in biofilters. *Bioresource Technology*, **100**, (17), 3869–3876.

Juntranapaporn, J., Vikromvarasiri, N., Soralump, C. and Pisutpaisal, N. (2019) Hydrogen sulphide removal from biogas in biotrickling filter systems inoculated with *Paracoccus Pantotrophus*. *International Journal of Hydrogen Energy*, **44**, (22), 29554–29560.

Kawase, Y., Hirata, A., Kojoma, T., Ohmori, S., Akutagawa, H., Uehara, K., Iwata, K., Nakajima, T. and Yamamoto, K. (2014) Improvement of biodegradation in compact co-current biotrickling filter by high recycle liquid flow rate: Performance and biodegradation. *Process Biochemistry*, **49**, (10), 1733–1740.

Kennes, C. and Veiga, M.C. (2004) Fungal biocatalysts in the biofiltration of VOC-polluted air. *Journal of Biotechnology*, **113**, 305–319.

La, H., Hettiaratchi. J.P., Achari, G. and Dunfield, P.F. (2018) Biofiltration of methane. *Bioresource Technology*, **268**, 759–772.

Lebrero, R., Estrada, J.M., Muñoz, R. and Quijano, G. (2014) Deterioration of organic packing materials commonly used in air biofiltration: Effect of VOC-packing interactions. *Journal of Environmental Management*, **137**, 93–100.

Lebrero, R., López, J.C., Lehtinen, I., Póérez, R., Quijano, G. and Muñoz, R. (2016) Exploring the potential of fungi for methane abatement: Performance evaluation of a fungal-bacterial biofilter. *Chemosphere*, **144**, 97–106.

Malakar, S., Saha, P.D., Baskaran, D. and Rajamanickam, R. (2017) Comparative study of biofiltration process for treatment of VOCs emission from petroleum refinery wastewater — a review. *Environmental Technology and Innovation*, **8**, 441–461.

Oliver, J.P., Janni, K.A. and Schilling, J.S. (2016) Bait and scrape: An approach for assessing biofilm communities on organic media used for gas-phase biofiltration. *Ecological Engineering*, **91**, 50–57.

Rybarczyk, P., Szulczyński, B., Gębicki, J. and Hupka, J. (2019) Treatment of malodorous air in biotrickling filters: A review. *Biochemical Engineering Journal*, **141**, 146–162.

Schiavon, M., Ragazzi, M., Rada, E.C. and Torretta, V. (2016) Air pollution control through biotrickling filters: A review considering operational aspects and expected performance. *Critical Reviews in Biotechnology*, **36**, 1143–1155. https://doi.org/10.3109/ 07388551.2015.1100586.

Van Groenestijn, J.W. (2001) Bioscrubbers. In *Bioreactors for Waste Gas Treatment. Environmental Pollution, Vol. 4* (eds. C. Kennes and M.C. Veiga). Springer, Dordrecht, pp. 133–162.

van Groenestijn, J.W., van Heiningen, W.N.M. and Kraakman, N.J.R. (2001) Biofilters based on the action of fungi. *Water Science and Technology*, **44**, (9), 227–232.

Vikrant, K., Kailasa, S.K., Tsang, D.C.W., Lee, S.S., Kumar, P., Giri, B.S., Singh, R.S. and Kim, K.-H. (2018) Biofiltration of hydrogen sulphide: Trends and challenges. *Journal of Cleaner Production*, **187**, 131–147.

Williams, T.O. and Miller, F.C. (1992a) Odour control using biofilters. *BioCycle*, **33**, (10), 72–77.

Williams, T.O. and Miller, F.C. (1992b) Biofilters and facilities operations. *BioCycle*, **33**, (11), 75–79.

Wu, J., Jiang, X., Jin, Z., Yang, S. and Zhang, J. (2020) The performance and microbial community in a slightly alkaline biotrickling filter for the removal of high concentration H2S from biogas. *Chemosphere*, **249**, 126–127.

Zhong, X.-Z., Sun, Z.Y., Wang, S.-P., Tang, Y.-Q., Kida, K. and Tanaka, A. (2020) Minimizing ammonia emissions from diary composting by biofiltration using a pre-composted material as the packing media. *Waste Management*, **102**, 569–578.

Zhou, Q., Zhang, L., Chen, J., Xu, B., Chu, G. and Chen, J. (2016) Performance and microbial analysis of two different inocula for the removal of chlorobenzene in biotrickling filters. *Chemical Engineering Journal*, **284**, 174–181. https://doi.org/10.1016/J.CEJ.2015.08.130.

# 9
# The Future for Fixed-Film Systems

Attached growth or fixed-film treatment systems are based on the natural process of self-purification found in every stream and river. As water flows downstream, it not only passes over the biofilm that has developed on surface stones, debris and plant surfaces but also through the layers of gravels over which the river flows. The surface biofilm is dominated by algae and diatoms as well as bacteria, but as the depth increases the biofilm becomes wholly bacterial and fungal in nature. The surface water is pulled down into this unexposed layer of gravel, known as the hyporheic zone, where colloidal and soluble organic matter is oxidized, and so too is ammonia to nitrate. The river water is constantly cycled through this natural fixed-film system, with the quality of the water improving accordingly. So fixed-film systems are integral in keeping our rivers clean. This natural process has been exploited by humans in many different ways. For example, biofiltration swales (bioswales) are used to treat road run-off, which is polluted by a wide range of trace organics, hydrocarbons and metals. These are wide shallow ditches covered with grass into which the run-off flows and is retained. The run-off slowly percolates into the biofilter below the grass sward, which is made up of engineered soil, usually, a high sand-loam mix that has a high permeability but good adsorption properties retaining contaminants and oxidizing any organics present (Flanagan *et al.*, 2018).

Although activated sludge is still the most widely used biological treatment process, it has a high carbon footprint, is complex to operate, and is often unreliable, especially when being used for joint carbonaceous and nutrient removal (Table 3.1). Fixed-film reactors (FFRs) have developed significantly from the basic percolating or trickling filter design to offer a range of highly sustainable and flexible treatment options. The key areas of development are:

- Better understanding of the formation and hence exploitation of biofilms in wastewater treatment,
- Moving toward closer control of the microbial ecology,
- Development of novel species reactors for N removal,
- More integrated use with other systems,
- Better designs and more closely controlled reactor environments,
- Sustainability of reactors in terms of longevity, energy use, GHG emissions, with the possibility of having zero-carbon systems, and
- Temporary use of packaged reactors in emergency situations.

Underpinning all FFRs is their potential for sustainable operation. At the most basic level, effective percolating filters can be constructed out of local or waste materials, which are frequently employed for rapid and temporary solutions for treatment for villages and small towns. For example, rushwood filters or filters made out of bamboo, which use the thicker canes to construct a framework that houses layers of thinner bamboo canes, which act as an effective medium (Kirchhof, 1989). Filters made out of old corrugated iron sheeting filled with plastic bottle tops are also equally effective. Percolating filter wastewater treatment works can be operated entirely by gravity and are exceptionally robust and long-lasting. Biofilms play an increasingly important role in the development of sustainable natural treatment systems such as grass filtration, slow and rapid infiltration, overland flow as well as in wetland systems (Gray, 2021).

FFRs are extremely flexible and new systems are constantly being developed. Recent examples include vermifiltration and photobiotic biofilm systems.

One of the most sustainable and cost-effective small scale filtration processes for domestic and industrial wastewaters is vermifiltration. Vermifilters are extremely simple in design and construction, making them ideal in poor and developing countries. They comprise a compost or bedding layer supported on a layer of quartz sand on top of a bottom layer of gravel, approximately 50% by volume of compost and 50% of sand and gravel. The wastewater is distributed onto the surface of the reactor with the final effluent discharged at the base. The treatment process is quite complex with the biofilm and the feeding activity of the earthworms (i.e., *Eisenia fetida*) being significant (Li *et al.*, 2013). It has removal efficiencies for domestic sewage of 84% of the biochemical oxygen demand (BOD), 81% of the chemical oxygen demand (COD), and 94% of the total suspended solids (TSS) (Wang *et al.*, 2016), with a high level of pathogen removal. While high organic loadings can be applied, it is intolerant of high hydraulic loadings. So, better performance is achieved when the filters are used in series (Lourenço and Nunes, 2017). Vermifiltration is also used in series with constructed wetlands (Samal *et al.*, 2017). The process has been reviewed by Singh *et al.* (2019), who have summarized the main removal mechanisms within vermifilters (Fig. 9.1).

Although FFRs are based almost entirely on heterotrophic and autotrophic bacteria, similar to those found in biofilms in the hyporheic zone of rivers, the surface films called the benthic periphyton has been largely ignored in terms of wastewater treatment potential. Algal and plants have always played an important role in natural treatment streams such as constructed wetland and algal ponds, although the direct use of phototrophic organisms as biofilms is an exciting new development.

Algal biofilm technology was developed in the southern part of the United States in the early 1990s, leading to the commercial treatment process Algal Turf Scrubber® (ATS™) manufactured by HydroMentia™. They comprise of growth mats or screens on which the algae develops, which can then be employed in many different scenarios. Algal growth on screens or nets has been used in rivers, lakes and even closed bays, where they strip out the nitrogen and phosphorus from the water and even help restore dissolved oxygen levels. The only energy requirement is sunlight, and they not only remove the nutrients but also remove carbon dioxide from the atmosphere, which makes them a very sustainable treatment option. ATS™ is also used for the tertiary treatment of wastewater. These units

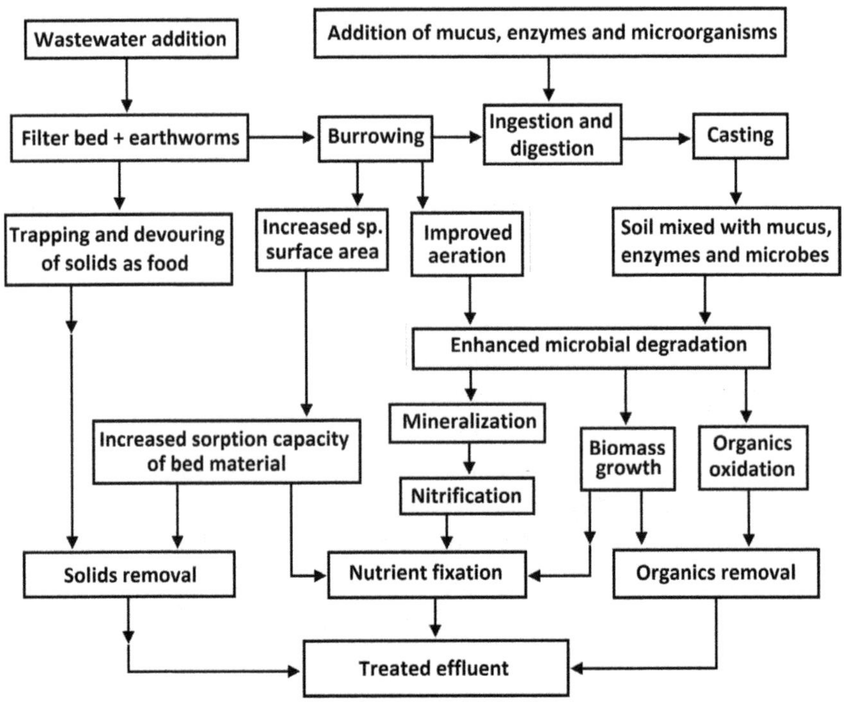


**Fig. 9.1.** Treatment mechanisms in a vermifilter (Singh *et al.*, 2019). Reproduced with permission of Elsevier Publishing.

require large, slightly inclined, textured surfaces (e.g., 7 m wide and 150 m in length), usually a flexible plastic material that is used as a growing surface for algae. The wastewater is pulsed over the surface, creating small waves that travel downward along the length of the unit, which encourages a luxuriant growth of periphytic algae that creates like what looks like a turf, from which the process gets its name. Removal rates of 1.1 and 0.7 g of TN [total nitrogen] and TP [total phosphorus] $m^{-2}d^{-1}$, respectively, can be achieved. The algae are regularly harvested and can be converted into energy (biodiesel or methane), livestock feed or a nutrient-rich compost (Adey *et al.*, 2011). These systems are effective but require a good sunny warm climate and sufficient available space.

Microalgal biofilms can also be grown in special photobioreactors that have been primarily used as a nutrient removal tertiary treatment process for domestic and municipal wastewaters (Boelee *et al.*, 2011). Microalgal biofilms are comprised of several layers, creating a complex community structure with photoautotrophs that include green algae and cyanobacteria, with all the expected heterotrophs such as bacteria, fungi and mesofauna. Like heterotrophic biofilms, microalgae can form a biofilm on almost any material, although the physical characteristics of the medium used affect the species of algae and cyanobacteria forming the community (Irving and Allen, 2011), with diatoms and green algal species colonizing first. The microalgal biofilm is similar in structure to bacterial biofilms, with it being held together by extracellular polymeric substances that form a highly porous matrix. Photobiotic filters are made up of horizontal or vertical panels over which the wastewater flows, and are designed to allow optimum exposure to either sun or artificial light. Their use for the treatment of settled wastewater is based on the symbiotic relationship between bacteria and the algae, which is the basis of the oxidation pond. The bacteria break down the organic matter releasing nutrients and carbon dioxide, while the algae utilize these nutrients and the carbon dioxide, which in turn releases oxygen that is subsequently used by the bacteria. However, this has proven difficult to achieve at full-scale using a biofilm reactor design. So their main function at the moment is for nutrient (N and P) removal. Sukačova and her colleagues used a photobioreactor to remove phosphorus from wastewater as a tertiary treatment step. They achieved a removal efficiency of 97% of the total phosphorus, equivalent to a removal rate of 0.16 g $m^{-2}d^{-1}$ when illuminated continuously (24 hours) using artificial light. This fell to between 36–41% when they used natural light (12 hours of light/12 hours of darkness) (Sukačova *et al.*, 2015).

Although currently produced in suspended cultures, the next generation of photobioreactors will look at utilizing the photosynthetic bacteria as the primary treatment organism. These include purple and green/non-green sulfur bacteria. These bacteria have been highly studied due to their ability to produce a wide range of valuable products, amongst which are polyhydroxyalkalanoates that are used for making bio-plastics, and the very valuable product 5-aminolevulinic acid, which is used in a wide range of pharmaceutical and chemical agricultural products. Photosynthetic bacteria can also be made into a single cell protein with an actual protein content of the bacteria between 40–90% (Cao *et al.*, 2020).

New developments in design and operation continue to be developed. For example, the rotating biological contactors (RBCs) used to treat sewage from rural villages in China have been made more efficient by returning a portion of the treated effluent and sludge into the influent (i.e., the New Integrated Self-Refluxing Rotating Biological Contactor (NISRRBC) process), thereby reducing sludge production and improving nitrification-denitrification (Han *et al.*, 2019). FFRs continue to be used in a number of novel ways to nitrify and denitrify wastewaters both in domestic and industrial settings as well as for phosphorous removal. For example, they are increasingly being used in the treatment of ammonia produced within closed circuit intensive fish farming operations for both freshwater and marine fish production (Eding *et al.*, 2006; Gutierrez-Wing and Malone, 2006). FFRs, including RBCs and fluidized beds, are increasingly being used in combination with advanced oxidation processes to treat toxic and non-biodegradable wastewaters (Oller *et al.*, 2011; Pariente *et al.*, 2013). Fluidized bed design and operation continues to develop and with new three-phase systems, such as aerobic inverse fluidized bed biofilm reactors (AIRFBBRs) (Swain *et al.*, 2018) and anaerobic fluidized bed reactors (AFBRs). These offer new opportunities for the treatment of contaminants that previously had to be treated by physico-chemical means, including gas desulfurization (Braz *et al.*, 2019; San-Valero *et al.*, 2019).

Using microbial supplements is increasingly common in wastewater technology. For example, the immobilization of white rot fungal biomass such *Trametes versicolor* and *Phanerochaete chrysosporium* to treat complex micropollutants such as Pharmaceutical Compounds (PhACs) has been successfully trialed at a pilot scale using a range of FFRs, including RBCs (Rodarte-Morales *et al.*, 2012; Cruz del Alamo *et al.*, 2020). White rot fungus has also been grown on cubes of corn cob, which are used as a carrier medium for use in FFRs. Apart from acting as the carrier, the cubes also act as a carbon source for the fungi as well as an absorbent for pollutants (Zhang *et al.*, 2017).

Discs in RBCs have undergone a constant transition to optimize biofilm development and stability, with radial corrugated plates being particularly popular. This transition has seen plates covered with polyurethane sponge or other material, although this usually results in heavy biofilm development, leading to a reduction in oxygen transfer and loss of overall performance. Rotating drum biological contactors are the next step in this development and are comprised of a porous drum filled with a random carrier or medium and can be used in series to create aerobic, anoxic and anaerobic units for biological nutrient removal (Šíma *et al.*, 2016; Li *et al.*, 2019). RBCs have been used for the treatment of a wide range of complex inorganic and organic wastewaters (Hassard *et al.*, 2015). For example, Padhi and Gokhale (2016) modified a standard RBC by replacing the discs with a sponge medium held in rotating drums to treat gaseous benzene.

In conclusion, FFRs are more robust, more sustainable, and able to deal better with fluctuating loads and shock events than activated sludge. Also, biofilms offer increasing opportunities to treat more complex and challenging wastewaters and contaminated gases. Due to advances in biochemistry, genetics and microbial ecology, the possibilities in the use of biofilms in wastewater treatment will continue to develop and expand.

## References

Adey, W.H., Kangas, P.C. and Mulbry, W. (2011) Algal turf scrubbing: Cleaning surface waters with solar energy while producing a biofuel. *Bioscience*, **61**, (6), 434–441.

Boelee, N.C., Temmink, H., Janssen, J., Buisman, C.J.N. and Wiffels, R.H. (2011) Nitrogen and phosphorus removal from municipal wastewater effluent using microalgal biofilms. *Water Research*, **45**, 5925–5933.

Braz, L.M., Aguiar, A.B.S., Rodriguez, R.P. and Sancinetti, G.P. (2019) Potential for anaerobic treatment of wastewater from PET bottle washing in a fluidized bed reactor. *Journal of Water Process Engineering*, **31**, 100817.

Cao, K., Zhi, R. and Zhang, G. (2020) Photosynthetic bacteria wastewater treatment with the production of value added products: A review. *Bioresource Technology*, **299**, 122648.

Cruz del Alamo, A., Pariente, M.I., Martinez, F. and Molina, R. (2020) *Trametes versicolor* immobilized on rotating biological contactors as alternative biological treatment for the removal of emerging concern micropollutants. *Water Research*, **170**, 115313. https://doi.org/10.1016/j.watres.2019.115313

Eding, E.H., Kamastra, A., Verreth, J.A.J., Huisman, E.A. and Klapwijk, A. (2006) Design and operation of nitrifying trickling filters in recirculating agquiculture: A review. *Aquacultural Engineering*, **34**, (3), 234–260.

Flanagan, K., Branchu, P. Boudahmane, L., Caupos, E., *et al.* (2018) Field performance of two biofiltration systems treating micro-pollutants from road runoff. *Water Research*, **145**, 562–578.

Gray, N.F. (2021) *Natural Wastewater Treatment Systems.* World Scientific Publishing, Singapore.

Gutierrez-Wing, M.T. and Malone, R.F. (2006) Biological filters in aquaculture: Trends and research directions for freshwater and marine applications. *Agricultural Engineering*, **34**, (3), 163–171.

Han, Y., Ma, J., Xiao, B., Huo, X. and Guo, X. (2019) New integrated self-refluxing rotating biological contactors for rural sewage treatment. *Journal of Cleaner Production*, **217**, 324–334.

Hassard, F., Biddle, J., Cartmell, E., Jefferson, B., Tyrrel, S. and Stephenson, T. (2015) Rotating biological contactors for wastewater treatment — a review. *Process Safety and Environmental Protection*, **94**, 285–306.

Irving, T.E. and Allen, D.G. (2011) Species and material considerations in the formation and development of microalgal biofilms. *Applied Microbiology and Biotechnology*, **92**, (2), 283–294.

Kirchhof, W. (1989) Developing ideas for suitable systems. *Water Quality International*, **1989**, (1), 20–21.

Li, N., Zeng, W., Yang, Y., Wang, B., Li, Z. and Peng, Y. (2019) Oxygen mass transfer and post-denitrification in a modified rotating drum biological contactor. *Biochemical Engineering Journal*, **144**, 48–56.

Li, X., Xing, M., Yang, J. and Lu, Y. (2013) Properties of biofilm in a vermifiltration system for domestic wastewater sludge stabilization. *Chemical Engineering Journal*, **223**, 932–943.

Lourenço, N. and Nunes, L.M. (2017) Optimization of a vermifiltration process for treating urban wastewater. *Ecological Engineering*, **100**, 138–146.

Oller, I., Malato, S. and Sánchez-Pérez, J.A. (2011) Combination of advanced oxidation processes and biological treatments for wastewater decontamination — a review. *Science of the Total Environment*, **409**, 4141–4166.

Padhi, S.K. and Gokhale, S. (2016) Benzene control from waste gas streams with a sponge-medium based rotating biological contactor. *International Biodeterioration and Biodegradation*, **109**, 96–103.

Pariente, M.I., Siles, J.A., Molina, R., Botas, J.A., Melero, J.A. and Martinez, F. (2013) Treatment of an agrochemical wastewater by integration of heterogenous catalytic wet hydrogen peroxide oxidation and rotating biological contactors. *Chemical Engineering Journal*, **226**, 409–415.

Rodarte-Morales, A.I., Feijoo, G., Moreira, M.T. and Lema J.M. (2012) Operation of stirred tank reactors (STRs) and fixed-bed reactors (FBRs) with free and immobilized *Phanerochaete chrysosporium* for the continuous removal of pharmaceutical compounds. *Biochemical Engineering Journal*, **66**, 38–45.

Samal, K., Roshan, R. and Bhunia, P. (2017) Treatment of wastewater by vermifiltration integrated with macrophyte filter: A review. *Journal of Environmental Chemical Engineering*, **5**, (3), 2274–2289.

San-Valero, P., Penya0roja, J.M., Álvarez-Hornos, F.J., Buitrón, G., Gabaldón, C. and Quijano, G. (2019) Fully aerobic bioscrubber for the desulfurization of $H_2S$-rich biogas. *Fuel*, **241**, 884–891.

Šíma, J., Pocedič, J. and Hasal, P. (2016) Decolorization of reactive orange 16 in rotating drum biological contactor. *Journal of Environmental and Chemical Engineering*, **4**, 4540–4548. https://doi.org/10.1016/j.jece.2016.10.010

Singh, R., Samal, K., Dash, R.R. and Bhunia, P. (2019) Vermifiltration as a sustainable natural treatment technology for the treatment and reuse of wastewater: A review. *Journal of Environmental Management*, **247**, 140–151.

Sukačova, K., Trtilek, M. and Rataj, T. (2015) Phosphorus removal using biofilm in a new biofilm photobioreactor for tertiary wastewater treatment. *Water Research*, **71**, 55–63.

Swain, A.K., Sahoo, A., Jena, H.M. and Patra, H. (2018) Industrial wastewater treatment by aerobic inverse fluidized bed biofilm reactors (AIFBBRs): A review. *Journal of Water Process Engineering*, **23**, 61–74.

Wang, Y., Xing, M.Y., Yang, J. and Lu, B. (2016) Addressing the role of earthworms in treating domestic wastewater by analyzing biofilm modification through chemical and spectroscopic methods. *Environmental Science and Pollution Research*, **23**, 4768–4777. https://doi.org/10. 1007/s11356-015-5661-6

Zhang, A., Wang, G., Gong, G. and Shen, J. (2017) Immobilization of white rot fungi to carbohydrate-rich corn cob as a basis for tertiary treatment of secondarily treated pulp and paper mill wastewater. *Industrial Crops and Products*, **109**, 538–541.

# About the Author

Professor Nick Gray is a founding member and former Director of the Trinity Centre for the Environment, a hub for interdisciplinary research and teaching at Trinity College Dublin. He is an expert hydrobiologist specializing in biological wastewater treatment and water pollution control. He combines his research and experience as a consultant environmental engineer and scientist into his teaching and writing. He has written over 180 papers in 49 different journals and is the author of 14 books and numerous book chapters.

# Index

Lightning Source UK Ltd.
Milton Keynes UK
UKHW021429130820
368175UK00001B/21